The Chemistry of Polymers

5th Edition

The Chemistry of Polymers

5th Edition

John W. Nicholson

Bluefield Centre for Biomaterials Ltd., London, UK

THE QUEEN'S AWARDS
FOR ENTERPRISE:
INTERNATIONAL TRADE
2013

Print ISBN: 978-1-78262-832-3

EPUB eISBN: 978-1-78801-114-3

A catalogue record for this book is available from the British Library

The Royal Society of Chemistry is a charity, registered in England and Wales, Number 207890, and a company incorporated in England by Royal Charter (Registered No. RC000524), registered office: Burlington House, Piccadilly, London W1J 0BA, UK, Telephone: +44 (0) 207 4378 6556.

Visit our website at www.rsc.org/books

Printed in the United Kingdom by CPI Group (UK) Ltd, Croydon, CR0 4YY, UK

Preface to First Edition

This book seeks to introduce readers to the chemistry of polymers. It is aimed primarily at new graduates who have not previously studied polymer chemistry as part of their degree course, but it ought to prove useful to others as well. I hope that final year students studying polymer chemistry and more experienced chemists in industry looking for concise information on the subject will find their needs met as well.

In preparing this book I have attempted to do two distinct things: (i) to provide a brief, readable introduction to the chemistry of polymers and (ii) to emphasise the applied aspects of the scientific knowledge presented. I believe that any introductory book ought to be written in a way that encourages its proposed audience actually to read it. Also as an applied scientist myself I want to emphasise the applications of polymer chemistry since I believe that these are worth covering in a book of this type.

No book, however brief, is the product of just one person. In my own case, I have been very conscious of the help I have received from many sources. These include authors of those books on polymer chemistry that I have drawn on in preparing this one, all of which are listed in the Bibliography.

I also want to give special thanks to my colleague Eleanor Wasson for her generous assistance in reading the entire manuscript and for her numerous helpful suggestions.

Lastly I thank my wife Suzette for her forbearance and support during my involvement with this project. Her good-natured tolerance of my frequent absence from family life has considerably assisted me in the completion of this book.

John Nicholson

The Chemistry of Polymers, 5th Edition
By John W. Nicholson
© John W. Nicholson 2017
Published by the Royal Society of Chemistry, www.rsc.org

Preface to Second Edition

The first edition of this book was, with one or two exceptions, gratifyingly well received by reviewers. It also sold well, thus confirming my belief that there is, indeed, place for an introductory book biased towards the more applied aspects of polymer science. It has been translated into Polish, suggesting that this need is international. Although it was not intended as a textbook for undergraduates, I have discovered that the first edition has been used for that purpose in certain universities, a point I have borne in mind in preparing the current edition.

In this second edition, I have stuck very much to the approach and level of the first edition, making changes only where absolutely necessary. In some cases this has been done to widen the coverage; in others it has been done to bring the text up to date. In making these changes, I have been helped by individuals who wrote to me after the publication of the first edition, with corrections or suggestions for additional material; I acknowledge my sincere gratitude for their help.

Despite this help, and the success of the first edition, I am sure that, even now, the book is not perfect. I am therefore still open to suggestions about how it could be improved. In the meantime, I take full responsibility for all the errors and infelicities that remain.

John Nicholson

The Chemistry of Polymers, 5th Edition
By John W. Nicholson
© John W. Nicholson 2017
Published by the Royal Society of Chemistry, www.rsc.org

Preface to Third Edition

The continuing success of this book has necessitated a further revision to bring it up to date. The main chapters have not been altered, but I have introduced new material into chapters 10 and 11, and I have also added a completely new chapter, number 9, to deal with the rapidly developing subject of dendrimers. Though what I have said here is, of necessity, brief it does draw attention to the existence of these materials and to the major ways in which they differ from conventional linear polymers.

I would like to take this opportunity to thank those readers who have written to me in the past to draw my attention to inaccuracies or ambiguities in the previous editions. May I invite readers of this new edition to consider doing the same? No book is perfect, but with your help, I can at least ensure that this one is the best it can be.

Lastly, once again I wish to acknowledge the support of my wife Suzette during the time I have devoted to the revision of this book. As ever, such an addition to my workload has only been possible because of her help, and I am deeply grateful to her for it.

John Nicholson
University of Greenwich at Medway

The Chemistry of Polymers, 5th Edition
By John W. Nicholson
© John W. Nicholson 2017
Published by the Royal Society of Chemistry, www.rsc.org

Preface to Fourth Edition

Once again, I have had the opportunity of revising and updating this book, for which I express my gratitude to the publishers. As in previous editions, the essential parts of the fundamental chapters have not been altered, but some of the later chapters have been updated. In particular, I have added brief introductory sections on the important topics of polymers for drug delivery and on "green" polymerisation. I have also added sections on the so-called RAFT and ATRP polymerisation processes, both of which are radical-based "living polymerisations" and which are of growing importance in the field of polymer synthesis.

As with the three prevision editions of this book, I acknowledge that carrying out the work of the current revision would not have been possible without the kind support of my wife Suzette, to whom I remain as indebted as ever.

John Nicholson
University of Greenwich at Medway

The Chemistry of Polymers, 5th Edition
By John W. Nicholson
© John W. Nicholson 2017
Published by the Royal Society of Chemistry, www.rsc.org

Preface to the Fifth Edition

I continue to be gratified by the success of this little book and was delighted when the publishers approached me to prepare a 5th edition. As before, I have updated key sections, for example those dealing with pollution by polymers and I have also added new sections, notably on Ring-Opening Methathesis Polymerisation and on MALDI mass spectrometry.

I appreciate the support of various people in preparing this book, in particular my wife Suzette, and also the various people who have commented on previous editions. Their remarks have prompted some of the changes and corrections in the current edition. With this help, I hope that this book continues to be a readable and useful introduction to the broad field of polymer chemistry.

John Nicholson
London

The Chemistry of Polymers, 5th Edition
By John W. Nicholson
© John W. Nicholson 2017
Published by the Royal Society of Chemistry, www.rsc.org

Contents

1	Polymer Chemistry		1
	1.1	Basic Concepts	1
	1.2	The History of the Concept of the Macromolecule	2
	1.3	Classification of Polymers	4
	1.4	Structure and Properties of Polymers	6

2	Polymerisation Reactions		25
	2.1	Chain Polymerisation	26
	2.2	Arrangement of Monomer Units	29
	2.3	Kinetics of Chain Polymerisation	30
	2.4	Autoacceleration	32
	2.5	Practical Methods of Chain Polymerisation	33
	2.6	Other Chain Polymerisation Mechanisms	37
	2.7	Recent Developments in "Living" Polymerisation	39
	2.8	Step Polymerisation	43
	2.9	Step Polymerisation with Polyfunctional Monomers	44
	2.10	Copolymerisation	45

3	Polymer Structure		48
	3.1	Polymer Stereochemistry	48

The Chemistry of Polymers, 5th Edition
By John W. Nicholson
© John W. Nicholson 2017
Published by the Royal Society of Chemistry, www.rsc.org

3.2   Polymer Crystallinity   50
3.3   Thermal and Mechanical Properties   54

4   Crosslinking   64

4.1   Introduction   64
4.2   Phenol–Formaldehyde Resins   67
4.3   Unsaturated Polyester Resins   69
4.4   Polyurethanes   72
4.5   Epoxy Resins   74

5   Polymer Solutions   77

5.1   Introduction   77
5.2   Dissolution of Polymers   77
5.3   Solubility Parameters   78
5.4   Simple Liquid Mixtures and Raoult's Law   80
5.5   Entropy of Mixing   81
5.6   Real Molecules in Dilute Solution   83
5.7   Shapes of Polymer Molecules in Solution   84
5.8   Reptation Model of Molecular Motion   86
5.9   Water-soluble Polymers   87
5.10   Uses of High-viscosity Polymer Solutions   89
5.11   Polymer Melts   90

6   Methods of Determining Relative Molar Mass   93

6.1   Introduction   93
6.2   Molar Masses from Colligative Properties   95
6.3   Light Scattering   97
6.4   Viscosity Methods of Determining Relative
       Molar Mass   102
6.5   End Group Analysis   104
6.6   Gel Permeation Chromatography   105
6.7   MALDI Mass Spectrometry   108

7 Mechanical Properties of Polymers 111

7.1 Introduction 111
7.2 Stress, Strain, and Young's Modulus 112
7.3 Brittle and Tough Fracture 112
7.4 Types of Strength 115
7.5 The Influence of Surfaces 116
7.6 Viscoelasticity 119
7.7 Creep and Stress Relaxation 121
7.8 Cold Drawing 123
7.9 Dynamic Tests 124
7.10 Time/Temperature Relationship 126
7.11 Rubberlike Elasticity 127
7.12 Reinforced Polymers 130
7.13 Practical Measurements of Mechanical Properties 133
Further Reading 135

8 Polymer Degradation 136

8.1 Introduction 136
8.2 Behaviour of Polymers in Fires 136
8.3 Assessment of Combustion Behaviour 138
8.4 Improvement of Stability of Polymers in Fires 139
8.5 Weathering of Polymers 141
8.6 Protection of Polymers From Photo-oxidation 143
8.7 Biological Degradation of Polymers 145
8.8 Exploitation of Polymer Degradation 147
8.9 Polymer Light Emitting Diodes 150

9 Dendrimers 152

9.1 Introduction 152
9.2 Organic Dendrimers 155
9.3 Supramolecular Dendrimers 158
9.4 Shapes and Conformations of Dendrimers 159
9.5 Dendrimer Micelles 160
9.6 Characterization of Dendrimers 161

9.7   Applications of Dendrimers   166
9.8   Megamers   168
      Further Reading   169

10   Special Topics in Polymer Chemistry   170

10.1   Introduction   170
10.2   Polymers in Medicine   170
10.3   Ionomers   173
10.4   Electronically Conducting Polymers   176
10.5   Interpenetrating Polymer Networks   178
10.6   Inorganic Polymers   180
10.7   Polymer Liquid Crystals   183
10.8   Polymers for Food Packaging   185
10.9   Polymers for Controlled Drug Delivery   187
10.10 Lithium–Polymer Batteries   190
10.11 Co-ordination Polymers   191
       Further Reading   192

11   Polymers and the Environment   193

11.1   Introduction   193
11.2   Pollution by Polymers   193
11.3   The Nature of the Problem   196
11.4   Polymers and Energy   198
11.5   Recycling of Polymers   198
11.6   Degradable Polymers   200
11.7   Green Polymerisation   201
11.8   The Future   202
       Further Reading   203

Bibliography   204

Subject Index   205

1 Polymer Chemistry

1.1 Basic Concepts

A polymer is a large molecule built up from numerous smaller molecules. These large molecules may be linear, slightly branched, or highly interconnected. In the latter case the structure develops into a large three-dimensional network.

The small molecules used as the basic building blocks for these large molecules are known as *monomers*. For example the commercially important material poly(vinyl chloride) is made from the monomer vinyl chloride. The repeat unit in the polymer usually corresponds to the monomer from which the polymer was made. There are exceptions to this, though. Poly(vinyl alcohol) is formally considered to be made up of vinyl alcohol (CH_2CHOH) repeat units but there is, in fact, no such monomer as vinyl alcohol. The appropriate molecular unit exists in the alternative tautomeric form, ethanal CH_3CHO. To make this polymer, it is necessary first to prepare poly(vinyl ethanoate) from the monomer vinyl ethanoate, and then to hydrolyse the product to yield the polymeric alcohol.

The size of a polymer molecule may be defined either by its mass (see Chapter 6) or by the number of repeat units in the molecule. This latter indicator of size is called the *degree of polymerisation*, DP. The relative molar mass of the polymer is thus the product of the relative molar mass of the repeat unit and the DP.

There is no clear cut boundary between polymer chemistry and the rest of chemistry. As a very rough guide, molecules of relative molar mass of at least 1000 or a DP of at least 100 are considered to fall into the domain of polymer chemistry.

The Chemistry of Polymers, 5th Edition
By John W. Nicholson
© John W. Nicholson 2017
Published by the Royal Society of Chemistry, www.rsc.org

The vast majority of polymers in commercial use are organic in nature, that is they are based on covalent compounds of carbon. This is also true of the silicones which, though based on silicon-oxygen backbones, also generally contain significant proportions of hydrocarbon groups. The other elements involved in polymer chemistry most commonly include hydrogen, oxygen, chlorine, fluorine, phosphorus, and sulfur, *i.e.* those elements which are able to form covalent bonds, albeit of some polarity, with carbon.

As is characteristic of covalent compounds, in addition to primary valence forces, polymer molecules are also subject to various secondary intermolecular forces. These include dipole forces between oppositely charged ends of polar bonds and dispersion forces which arise due to perturbations of the electron clouds about individual atoms within the polymer molecule. Hydrogen bonding, which arises from the particularly intense dipoles associated with hydrogen atoms attached to electronegative elements such as oxygen or nitrogen, is important in certain polymers, notably proteins. Hydrogen bonds have the effect of fixing the molecule in a particular orientation. These fixed structures are essential for the specific functions that proteins have in the biochemical processes of life.

1.2 The History of the Concept of the Macromolecule

Modern books about polymer chemistry explain that the word polymer is derived from the Greek words 'poly' meaning many and 'meros' meaning part. They often then infer that it follows that this term applies to giant molecules built up of large numbers of interconnected monomer units. In fact this is misleading since historically the word polymer was coined for other reasons. The concept of polymerism was originally applied to the situation in which molecules had identical empirical formulae but very different chemical and physical properties. For example, benzene (C_6H_6; empirical formula CH) was considered to be a polymer of acetylene (C_2H_2; empirical formula also CH). Thus the word 'polymer' is to be found in textbooks of organic chemistry published up to about 1920 but not with its modern meaning.

The situation is confused, however, by the case of certain chemicals. Styrene, for example, was known from the mid-nineteenth century as a clear organic liquid of characteristic pungent odour. It was also known to convert itself under certain circumstances into a clear resinous solid that was almost odour-free, this resin then being called metastyrene. The formation of metastyrene from styrene was described as a polymerisation and metastyrene was held to be a polymer of styrene. However, these terms applied only in the sense that there was no change in empirical formula despite the very profound alteration in chemical and physical properties. There was no understanding of the cause of this change and certainly the chemists of the time had no idea of what had happened to the styrene that was remotely akin to the modern view of polymerisation.

Understanding of the fundamental nature of those materials now called polymers had to wait until the 1920s, when Herman Staudinger coined the word 'macromolecule' and thus clarified thinking. There was no ambiguity about this new term – it meant 'large molecule', again from the Greek, and these days is used almost interchangeably with the word polymer. Strictly speaking, though, the words are not synonymous. There is no reason in principle for a macromolecule to be composed of *repeating* structural units; in practice, however, they usually are. Staudinger's concept of macromolecules was not at all well received at first. His wife once recalled that he had 'encountered opposition in all his lectures'. Typical of this opposition was that of one distinguished organic chemist who declared that it was as if zoologists 'were told that somewhere in Africa an elephant was found who was 1500 feet long and 300 feet high'.

There were essentially three reasons for this opposition. Firstly, many macromolecular compounds in solution behave as colloids. Hence they were assumed to be identical with the then known inorganic colloids. This in turn implied that they were not macromolecular at all, but were actually composed of small molecules bound together by ill-defined secondary forces. Such thinking led the German chemist C. D. Harries to pursue the search for the 'rubber molecule' in the early years of the twentieth century. He used various mild degradations of natural rubber, which he believed would destroy the colloidal character of the material and yield its constituent molecules, which were assumed to be fairly small. He was, of course, unsuccessful.

The second reason for opposition to Staudinger's hypothesis was that it meant the loss of the concept of a single formula for a single compound. Macromolecules had to be written in the form $(CH_2CHX)_n$, where n was a large number. Moreover, no means were available, or indeed are available, for discretely separating molecules where $n = 100$ from those where $n = 101$. Any such attempted fractionation always gives a distribution of values of n and, even if the mean value of a fraction is actually $n = 100$, there are significant numbers of molecules of $n = 99$, $n = 101$, and so on. Now the concept of one compound, one formula, with one formula being capable of both physical (*i.e.* spatial) and chemical interpretation, had been developed slowly and at some cost, with many long, hard-fought battles. Organic chemists could not easily throw it out, particularly in view of the fact that it had been so conspicuously successful with much of the rest of organic chemistry.

The third reason for opposition lay in the nature of many of the polymeric materials then known. Not only were they apparently ill-characterised, but they were also frequently non-crystalline, existing as gums and resins. Just the sort of unpromising media, in fact, from which dextrous organic chemists had become used to extracting crystalline substances of well characterised physical and chemical properties. To accept such resins as inherently non-crystallisable and not capable of purification in the traditional sense of the word was too much for the self-esteem of many professional organic chemists.

Staudinger's original paper opposing the prevalent colloidal view of certain organic materials was published in 1920 and contained mainly negative evidence. Firstly, he showed that the organic substances retained their colloidal nature in all solvents in which they dissolve; by contrast, inorganic colloids lose their colloidal character when the solvent is changed. Secondly, contrary to what would have been expected, colloidal character was able to survive chemical modification of the original substance.

By about 1930 Staudinger and others had accumulated much evidence in favour of the macromolecular hypothesis. The final part in establishing the concept was carried out by Wallace Carothers of the Du Pont company in the USA. He began his work in 1929 and stated at the outset that the aim was to prepare polymers of definite structure through the use of established organic reactions. Though his personal life was tragic, Carothers was an excellent chemist who succeeded brilliantly in his aim. By the end of his work he had not only demonstrated the relationship between structure and properties for a number of polymers, but he had invented materials of tremendous commercial importance, including neoprene rubber and the nylons.

1.3 Classification of Polymers

There are a number of methods of classifying polymers. One is to adopt the approach of using their response to thermal treatment and to divide them into thermoplastics and thermosets. Thermoplastics are polymers which melt when heated and resolidify when cooled, while thermosets are those which do not melt when heated but, at sufficiently high temperatures, decompose irreversibly. This system has the benefit that there is a useful chemical distinction between the two groups. Thermoplastics comprise essentially linear or lightly branched polymer molecules, while thermosets are substantially crosslinked materials, consisting of an extensive three-dimensional network of covalent chemical bonding.

Another classification system, first suggested by Carothers in 1929, is based on the nature of the chemical reactions employed in the polymerisation. Here the two major groups are the *condensation* and the *addition* polymers. Condensation polymers are those prepared from monomers where reaction is accompanied by the loss of a small molecule, usually of water, for example polyesters which are formed by the condensation shown in Reaction (1.1).

$$n \text{ HO}-R-OH \ + \ n \text{ HOOC}-R^1-COOH \longrightarrow$$

$$\text{HO}[-R-COO-R^1-COO-]_n H \ + \ (n-1)H_2O \tag{1.1}$$

By contrast, addition polymers are those formed by the addition reaction of an unsaturated monomer, such as takes place in the polymerisation of vinyl chloride (Reaction (1.2)).

$$n \text{ CH}_2{=}CHCl \longrightarrow [-CH_2\text{-}CHCl\text{-}]_n \tag{1.2}$$

This system was slightly modified by P. J. Flory, who placed the emphasis on the mechanisms of the polymerisation reactions. He reclassified polymerisations as *step* reactions or *chain* reactions corresponding approximately to condensation or addition in Carother's scheme, but not completely. A notable exception occurs with the synthesis of polyurethanes, which are formed by reaction of iso-cyanates with hydroxy compounds and follow 'step' kinetics, but without the elimination of a small molecule from the respective units (Reaction (1.3)).

$$n \text{ OCN}-R-NCO \ + \ n \text{ HO}-R^1\text{-}OH \longrightarrow$$

$$\text{OCN}-[R\text{-}NHCO_2-R^1-]_n \text{ OH} \tag{1.3}$$

In the first of these, the kinetics are such that there is a gradual build up of high relative molar mass material as reaction proceeds, with the highest molar mass molecules not appearing until the very end of the reaction. On the other hand, chain reactions, which occur only at a relatively few activated sites within the reacting medium, occur with rapid build up of a few high relative molar mass molecules while the rest of the monomer remains unreacted. When formed, such macromolecules stay essentially unchanged while the rest of the monomer undergoes conversion. This means that large molecules appear very early in the polymerisation reaction, which is characterised by having both high relative molar mass and monomer molecules present for most of the duration of the reaction.

Step reactions can give molecules having a variety of morphologies from the simple unbranched linear to the heavily crosslinked network. The final structure depends on the number of functional groups in the parent monomers – the greater the proportion with a functionality of greater than two, the more extensive will be the branching until, at sufficient degrees of branching, a highly cross-linked network emerges. Chain reactions, by contrast, give only linear or lightly branched polymers. Thus, in terms of the thermoplastic/thermoset classification, chain reactions give thermoplastics, while step reactions may give either thermoplastics or thermosets.

1.4   Structure and Properties of Polymers

Having established the basic principles of classification in polymer chemistry, we will now turn our attention to individual polymers and consider a little about their physical and chemical properties. Most of the examples which follow are of commercial importance, though it is their properties that are emphasised rather than the complete chemistry of their industrial manufacture. In terms of the classification systems just referred to, it is found that the polymers of major commercial importance tend to be the thermoplastics or addition polymers rather than the thermosets, and these will be dealt with first.

1.4.1   Poly(ethylene)

This polymer has one of the simplest molecular structures ($[CH_2CH_2-]_n$) and is at present the largest tonnage plastic material, having first been produced commercially in 1939 for use in electrical insulation. There is a difficulty over the nomenclature of this polymer. The IUPAC recommended name for the monomer is ethene, rather than the older ethylene. Hence the IUPAC name for the polymer is poly (ethene). However, this name is almost never used by chemists working with the material; throughout this book, therefore, this polymer will be referred to by its more widespread name, poly(ethylene).

There are four different industrial routes to the preparation of poly (ethylene), and they yield products having slightly different properties. The four routes are:

(i) High Pressure Processes,
(ii) Ziegler Processes,
(iii) The Phillips Process,
(iv) The Standard Oil (Indiana) Process.

The first group of these use pressures of 1000–3000 atm and temperatures of 80–300 °C. Free-radical initiators, such as benzoyl peroxide or oxygen, are generally used (see Chapter 2), and conditions need to be carefully controlled to prevent a runaway reaction, which would generate hydrogen, methane, and graphite rather than polymer. In general, high-pressure processes tend to yield lower density poly(ethylenes), typically in the range 0.915–0.945 $g\,cm^{-3}$, which also have relatively low molar masses.

Ziegler processes are based on co-ordination reactions catalysed by metal alkyl systems. Such reactions were discovered by Karl Ziegler in Germany and developed by G. Natta at Milan in the early 1950s.

A typical Ziegler–Natta catalyst is the complex prepared from titanium tetrachloride and triethylaluminium. It is fed into the reaction vessel first, after which ethylene is added. Reaction is carried out at low pressures and low temperatures, typically no more than 70 °C, with rigorous exclusion of air and moisture, which would destroy the catalyst. The poly(ethylenes) produced by such processes are of intermediate density, giving values of about 0.945 $g\,cm^{-3}$. A range of relative molar masses may be obtained for such polymers by varying the ratio of the catalyst components or by introducing a small amount of hydrogen into the reaction vessel.

Lastly the Phillips and the Standard Oil (Indiana) Processes both yield high density poly(ethylenes), using relatively low pressures and temperatures. Details of these processes are given in Table 1.1.

The polymer is a very familiar material in the modern world. It is a waxy solid which is relatively low in cost, easily processed, and shows good chemical resistance. Low relative molar mass grades have the disadvantage that they suffer from so-called 'environmental stress

Table 1.1   Details of processes for preparing poly(ethylene).

Process	Catalyst	Pressure/ atm	Temperature/ °C	Density of product/ $g\,cm^{-3}$
Philips	5% CrO_3 in finely divided silica/ alumina	15–35	130–160	9.6
Standard oil (Indiana)	Supported MoO_3 with Na, Ca metal or hydride promoters	40–80	230–270	9.6

cracking', that is they suddenly fail catastro-phically for no apparent reason after exposure to sunlight and moisture. Despite this draw-back, the various grades of poly(ethylene) have a wide range of uses. These include pipes, packaging, components for chemical plant, crates, and items for electrical insulation.

1.4.2   Poly(propylene)

This polymer, which has the structure $[-CH_2CH(CH_3)-]_n$ arose as a commercial material following the work of Natta on catalysts for the preparation of high relative molar mass polymers from alkenes. Following his work on the polymerisation of ethylene, Natta showed in 1954 that it was possible to prepare analogous polymers of propylene. Commercial exploitation followed rapidly, and poly (propylene) was first marketed in 1957.

Like poly(ethylene), there are formal problems with the nomenclature of this polymer, since its IUPAC name, poly(propene), is also rarely if ever used by polymer chemists. Since, in practice, no ambiguity is associated with the non-systematic name, this is the one that is generally used, as it will be throughout this book.

When poly(propylene) was first made, it was found to exist in two possible forms. One was similar to poly(ethylene), but had greater rigidity and hardness; the other was found to be amorphous and of little strength. The first of these is now known to be isotactic, that is with a regular stereochemistry at each alternating carbon atom. The other is now known to be atactic, that is with a random distribution of different stereochemical arrangements at each methyl-bearing carbon atom. This whole topic of polymer stereochemistry is dealt with in detail in Chapter 3.

Commercial poly(propylene) is usually about 90–95% isotactic and is very similar to poly(ethylene) with the following exceptions:

(i) It has a lower density ($0.90 \, \mathrm{g \, cm^{-3}}$).
(ii) It has a higher softening point and hence can be used at higher temperatures. For example, it is used to make kettles, generally of the jug-type of design, and it is found to be well able to withstand the effects of exposure to boiling water.
(iii) It is not susceptible to environmental stress cracking.
(iv) It is more readily oxidised. This is a consequence of the relatively readily fragmented tertiary C–H bond within the molecule. Another interesting property of isotactic poly(propylene) is that it can be repeatedly flexed without causing weakness. Hence it is used for one-piece mouldings, such as boxes for card indexes, which can be used over many years without damage.

1.4.3 Poly(methyl methacrylate)

This polymer, whose name is often abbreviated to PMMA, is the most important of the commercial *acrylic* polymers. These polymers are formally derived from poly(acrylic acid), $[-CH_2CH(CO_2H)-]_n$. In the case of PMMA, this derivation can be thought of as coming about by replacement of the tertiary hydrogen atom by a methyl group, CH_3, and by esterification of the carboxylic acid group with methanol, CH_3OH. This gives the structure (1.1).

$$[-CH_2-\underset{\underset{CO_2CH_3}{|}}{\overset{\overset{CH_3}{|}}{C}}-]_n$$

(1.1)

As with other major tonnage polymers, the IUPAC-recommended name for PMMA is almost never used, partly no doubt because it is extremely cumbersome. It is poly[l-(methoxycarbonyl)-l-methylethene].

Commercially PMMA is becoming relatively less important as time passes. For example, in the mid-1960s sales of PMMA ran at 40% of the total level of sales of poly(styrene); by 1980, they had fallen to 12%. PMMA does, though, retain certain uses, including interestingly as the material from which plastic dentures are manufactured.

PMMA is a transparent, glassy material, which is known from X-ray studies to be amorphous at the molecular level. It is clearly polar, as is shown by its poor insulating properties relative to poly(ethylene) and its solubility in solvents such as ethyl propanoate and trichloroethane.

The main applications for PMMA arise from the combination of its transparency and its good outdoor weathering properties. These are coupled with reasonable toughness and rigidity, so that PMMA is a useful material in a range of glazing applications. For example, it is used for display signs, street lamp fittings, and ceiling lights in factories and offices. It is also the standard material for automobile rear lamp housings.

1.4.4   Poly(styrene)

Poly(styrene) is a polymer which finds widespread use in the developed world on account of its desirable properties, combined with its relative cheapness. Among its features are excellent colour range, transparency, rigidity, and low water absorption.

The monomer, styrene, is a derivative of benzene, vinyl benzene (1.2). It is a colourless, mobile liquid that polymerises readily. The first report of the polymerisation reaction came in 1839, when E. Simon described the transformation of what was then called 'styrol'. He believed he had oxidised the material and called the product styrol oxide. Later, when it was realised that it contained no oxygen, the product became known as metastyrene.

$$CH_2{=}CH-\!\!\!\!\raisebox{-0.5ex}{\Large\text{⟨◯⟩}}$$

(1.2)

The polymer is based on a simple head-to-tail arrangement of monomer units and is amorphous, since the specific position of the benzene ring is somewhat variable and hence inhibits crystallisation. Despite its generally desirable properties, for many applications it is considered too brittle. Because of this, a number of approaches have been made to modify the mechanical properties of poly(styrene). The most successful of these have been (i) copolymerisation and (ii) the addition of rubbery fillers.

A very large number of different copolymers of styrene have been produced and sold commercially. Among the most successful have been those polymers also containing butadiene. At the appropriate ratio of monomers, these copolymers behave as synthetic elastomers, known as styrene–butadiene rubber, SBR. Another important group of materials are the terpolymers of acrylonitrile, butadiene, and styrene, ABS. A large range of ABS materials can be made by varying not only the ratios of the three monomers but also the precise conditions of manufacture. ABS materials generally have high impact strength and can be readily moulded to give articles of good surface finish.

The alternative approach to producing strengthened styrene polymers is to use rubbery fillers. Such materials are known as high-impact poly(styrenes), a name which reflects their improved mechanical properties. They are typically prepared by dissolving the rubber in the styrene monomer, followed by polymerisation of the styrene in the usual way. The process gives a blend containing not only rubber and poly(styrene), but also small amounts of graft copolymer, comprising short poly(styrene) side chains attached to the molecules of the rubber. The overall effect of using this approach is to give materials with much better impact strengths than pure poly(styrene).

Lastly, the polymerisation of styrene is used for crosslinking unsaturated polyesters. Typically a polyester is produced which contains double bonds. Such a resin, known as a pre-polymer, is then mixed with styrene to give a viscous solution, and this solution is then worked into a glass fibre mat. The resulting composite material develops its final properties by polymerisation of the styrene, which essentially undergoes homopolymerisation, but also incorporates the occasional double bond from the pre-polymer molecule. This gives a large, three-dimensional network, intertwined with the glass reinforcement. The composite material has excellent properties and is widely used for boat hulls, shower units, and baths.

1.4.5   Poly(vinyl chloride), PVC

Poly(vinyl chloride) is the widely accepted trivial name for poly-(1-chloroethene) and, in terms of worldwide production, is one of the three most important polymers in current use, the other two being poly(ethylene) and poly(styrene). PVC has numerous uses, including cable insulation, packaging, and toys.

PVC has been shown to have a head-to-tail structure. Typical experimental evidence for this is that when dissolved in dioxan and treated with zinc dust, it undergoes a Wurtz-type reaction to yield a product containing a small amount of chlorine and no detectable unsaturation. The alternative possible structure, the head-to-head arrangement, would yield unsaturated sites where adjacent chlorine atoms had been removed (Reaction (1.4)).

$$-CH_2-CH-CH-CH_2- \quad \xrightarrow{\ Zn\ } \quad -CH_2CH{=}CH-CH_2- \qquad (1.4)$$
$$\underset{Cl}{|}\ \ \underset{Cl}{|}$$

Uncompounded PVC is colourless and rigid, and possesses poor stability towards heat and light. Indeed, PVC is certainly the least stable of the high tonnage commercial polymers. Exposure to either light or heat at temperatures well below the softening point brings about a number of undesirable changes. Initially discoloration is apparent, as the polymer becomes yellow and passes progressively through deep amber to black on increasing time of exposure. At the same time the material becomes increasingly brittle.

These changes are caused by the polymer undergoing dehydro-chlorination, *i.e.* loss of HCl from along the backbone. Such loss is autocatalytic, and may continue until there are only traces of chlorine left in the macromolecule. It has the effect of reducing molecular flexibility and of allowing absorption of light at the visible part of the electromagnetic spectrum. These alterations in the molecular features lead to the observed changes in the properties of the bulk polymer.

In order to fabricate articles from PVC, and then use them in locations where they will be exposed to sunlight, it is necessary to add stabilisers at the compounding stage. A large number of chemicals have been used as stabilisers for PVC, including lead soaps, barium and cadmium laurates, and organotin compounds such as tin di-iso-octylthioglycollate. These latter compounds, the organosulfurtins, are relatively expensive but are used where good clarity and freedom from colour are important in the final compounded PVC.

The mechanisms of degradation and the mode of action of the various PVC stabilisers have both been widely studied. Often at least one aspect of their operation is some sort of reaction with the first trace of hydrogen chloride evolved. This removes what would other-wise act as the catalyst for further dehydrochlorination, and hence significantly retards the degradation process. In addition, many stabilisers are themselves capable of reacting across any double bonds formed, thereby reversing the process that causes discoloration and embrittlement.

1.4.6 The Nylons

The name Nylon was given by the Du Pont company of America to their first synthetic condensation polymer formed by the reaction of difunctional acids with difunctional amines. It had been made as part of the fundamental programme of W. H. Carothers to investigate the whole topic of polymerisation. The term has gradually been extended to other related polymers. These materials are strictly polyamides, but this term includes that otherwise distinct class of natural macro-molecules, the proteins. The term nylon is retained for its usefulness in distinguishing synthetic polyamides from the broader class of such polymers.

As mentioned a moment ago, nylons are condensation or step polymers and, because of this, they are different from all of the other commercially important polymers described so far. The nylons are distinguished from each other by a numbering system based on the number of carbon atoms in the starting materials. Thus, nylon 6,6, which was first prepared in 1935 and is still the major commercial nylon, is prepared by reaction of hexamethylenediamine (six carbon atoms) with adipic acid (six carbon atoms) (Reaction (1.5)).

$$n\,H_2N(CH_2)_6NH_2 \quad + \quad n\,HOOC(CH_2)_4COOH \longrightarrow$$
$$HN_2[(CH_2)_6NHCO(CH_2)_4CO]_n \ + \ (n-1)H_2O$$

$$(1.5)$$

Similarly nylon 6,10 is prepared from hexamethylenediamine and sebacic acid (ten carbon atoms). Certain nylons are designated not by two numbers, but by a single number; these are the nylons prepared from a single starting material. Such a starting material can be either a long-chain amino acid, such as ω-aminodecanoic acid (eleven carbon atoms) which undergoes self-condensation to yield nylon 11, or a closed ring amide-type compound, known as a lactam, which will undergo a ring-opening reaction to yield a polymer. This is the method by which nylon 6 is prepared from caprolactam (six carbon atoms) and nylon 12 is prepared from dodecyl-lactam (twelve carbon atoms). The latter reaction may be represented by the equation shown in Reaction (1.6).

$$n\,(CH_2)_{11}\underset{NH}{\overset{CO}{\Big|}} \longrightarrow [-(CH_2)_{11}CONH-]_n \qquad (1.6)$$
Nylon 12

The various nylons tend to have similar physical properties. These include high impact strength, toughness, flexibility, and abrasion resistance. Also, because of the linear structure, they form excellent fibres. This was the basis of the early commercial success of nylon 6,6. Commercial production of fibres of nylon 6,6 began in 1938, and two years later nylon stockings became generally available in the USA. As is well known, they were a huge success, and became highly fashionable during the years of World War II.

Nylons tend to show very good resistance to organic solvents, and also to fuels and oils. They are, however, readily attacked by concentrated mineral acids at room temperature and by alkalis at elevated temperatures.

Nylons 6,6 and 6 are the ones usually employed as textile fibres. Where individual monofilaments are used, such as in brushes, sports equipment, or surgical sutures, nylons 6,10 and 11 tend to be used, because of their greater flexibility and water resistance. All nylons can be injection moulded and the resulting articles have found widespread use in engineering applications, such as bearings and gears.

1.4.7 Epoxy Resins

These polymers are based on the three-membered heterocyclic system either as the epoxy or oxirane ring (1.3).

$$CH_2-CH-$$

(1.3)

For commercial application, diepoxides such as those derived from bisphenol A are employed, and they are cured *via* ring-opening crosslinking reactions, into which the epoxy group enters readily. Bisphenol A is so-called because it is formed from two moles of phenol and acetone (Reaction (1.7)).

$$2\ HO-\langle\ \rangle\ +\ (CH_3)_2C{=}O\ \longrightarrow$$

$$HO-\langle\ \rangle-\underset{CH_3}{\overset{CH_3}{C}}-\langle\ \rangle-OH\ +\ H_2O \qquad (1.7)$$

A variety of reagents can bring about these ring openings, including amines and anhydrides used in stoichiometric amounts, or Lewis acids such as $SnCl_4$ and Lewis bases such as tertiary amines, used in catalytic amounts.

The crosslinking reactions are illustrated in Reaction (1.8), and they demonstrate that, in principle, only a trace of curing agent is necessary to bring about cure of epoxy resins. Selection of curing agent depends on various considerations, such as cost, ease of handling, pot life, cure rates, and the mechanical, electrical, or thermal properties required in the final resin.

Epoxy resins are relatively expensive, but despite this, they are firmly established in a number of important applications. These include adhesives, protective coatings, laminates, and a variety of uses in building and construction.

$$B^- + H_2C-CH- \longrightarrow B-CH_2-\overset{O^-}{\underset{|}{CH}}- \xrightarrow{H_2C-CH-} \qquad (1.8)$$

$$\begin{array}{c} O^- \\ | \\ O-CH_2CH- \\ | \\ B-CH_2-CH- \end{array}$$

1.4.8 Phenol–Formaldehyde Polymers

These are thermoset polymers made from phenol or, less commonly, phenolic-type compounds such as the cresols, xylenols, and resorcinol, together with formaldehyde. They had been known for some time – G.T. (later Sir Gilbert) Morgan discovered them in the early 1890s when attempting (unsuccessfully) to make artificial dyestuffs by reaction of phenol with formaldehyde. But this knowledge had not been exploited before 1907, the year in which Leo Baekeland in America obtained his first patent for materials prepared from these two compounds. In 1910 he founded the General Bakelite Company to exploit this development, in the process making phenol–formaldehydes, the first synthetic polymers to achieve commercial importance.

Baekeland had to make important discoveries before he could bridge the gap between the initial concept and final products. In particular, he found that the base-catalysed condensation of phenol and formaldehyde can be carried out in two parts. If the process is carefully controlled, an intermediate product can be isolated, either as a liquid or a solid, depending on the extent of reaction. At this stage, the material consists of essentially linear molecules and is both fusible and soluble in appropriate solvents. When heated under pressure to 150 °C, this intermediate is converted to the hard, infusible solid known as 'bakelite'. This second stage is the one at which the three-dimensional cross-linked network develops.

Phenol–formaldehydes may no longer hold the centre-stage where synthetic polymers are concerned, but they are still of some commercial importance. They are produced for electrical mouldings, appliance handles, household fittings, and also as adhesives and specialised surface coatings.

Cured phenol–formaldehydes are resistant to attack by most chemicals. Organic solvents and water have no effect on them, though they will swell in boiling phenols. Simple resins are readily attacked by sodium hydroxide solutions, but resins based on phenol derivatives, such as cresol, tend to be less affected by such solutions. Simple phenol–formaldehyde polymers are resistant to most acids, though formic and nitric acids will tend to attack them. Again, cresol-based polymers have resistance to such attack.

1.4.9 Amino Resins

Amino resins are those polymers prepared by reaction of either urea or melamine with formaldehyde. In both cases the product that results from the reaction has a well crosslinked network structure, and hence is a thermoset polymer. The structures of the two parent amino compounds are shown in Figure 1.1.

The initial step of the polymerisation process is reaction of the amine groups with formaldehyde to generate methylol units, as illustrated in Reaction (1.9). Further heating of the polymer then leads to a variety of reactions. For example, the methylol groups can undergo self-condensation (Reaction (1.10)).

$$-NH_2 + HCHO \longrightarrow -NHCH_2OH \qquad (1.9)$$

$$2 \ -NHCH_2OH \longrightarrow -NHCH_2OCH_2NH- + H_2O \qquad (1.10)$$

Alternatively the methylol groups can react with further amino groups, also evolving a molecule of water in what is another condensation reaction (Reaction (1.11)).

$$-NHCH_2OH + H_2N- \longrightarrow -NHCH_2NH- + H_2O \qquad (1.11)$$

Unlike phenol–formaldehyde polymers, the amino resins are not themselves deeply coloured, but are of a naturally light appearance.

Urea   Melamine

Figure 1.1   Structures of urea and melamine.

They can be easily pigmented to give a variety of shades, which leads to application in uses where good appearance is highly valued, for example in decorative tableware, laminated resins for furniture, and modern white electrical plugs and sockets.

When crosslinked, amino resins are very resistant to most organic solvents, though they tend to be attacked by both acids and alkalis. Urea–formaldehyde polymers are more susceptible to attack than those prepared from melamine and formaldehyde.

1.4.10   Poly(tetrafluoroethylene), PTFE

This polymer is the completely fluorine-substituted analogue of poly(ethylene) *i.e.* $[CF_2CF_2-]_n$. The amount of this polymer produced commercially is very small compared with the output of many other synthetic polymers, but it has a number of important specialised uses and so is worth considering briefly.

PTFE is a white solid with a waxy appearance. It is very tough and flexible, with a good electrical insulation properties. The surface energy and coefficient of friction are both very low, the latter being the lowest of any solid. The combination of low surface energy and low coefficient of friction cause PTFE to have excellent non-stick characteristics, a feature which underlies many of the everyday uses of this polymer.

In addition to its unique mechanical properties, PTFE has excellent chemical resistance to a very wide range of reagents, with the exception of molten alkali metals and fluorine. There is no known solvent for PTFE.

PTFE is a linear polymer of the addition type, formed by polymerisation of the unsaturated monomer tetrafluoroethylene, $CF_2=CF_2$. Despite the fact that this structure ought to impart thermoplastic character to the polymer, PTFE does not show conventional melting behaviour. It does not apparently liquify on heating, nor does it give a melt that will flow. Instead it forms a high-viscosity translucent material which fractures rather than flows when an appropriate force is applied.

Because PTFE cannot be dissolved or melted, there are problems with fabricating articles out of the polymer. These are overcome by using a technique similar to powder metallurgy, in which granular particles are fused under high pressures and temperatures.

The main domestic use for PTFE is on non-stick utensils such as frying pans. Industrially, the polymer is used for gaskets, pump parts, and laboratory equipment.

1.4.11 Polyurethanes

'Urethane' is the name given to the functional group formed from the reaction of an isocyanate group with a hydroxy-group (Reaction (1.12)).

$$\text{—N=C=O} \quad + \quad \text{HO—} \quad \longrightarrow \quad \text{—NHCOO—} \tag{1.12}$$

Polyurethanes are thermoset polymers formed from di-isocyanates and polyfunctional compounds containing numerous hydroxy-groups. Typically the starting materials are themselves polymeric, but comprise relatively few monomer units in the molecule. Low relative molar mass species of this kind are known generally as *oligomers*. Typical oligomers for the preparation of polyurethanes are polyesters and polyethers. These are usually prepared to include a small proportion of monomeric trifunctional hydroxy compounds, such as trimethylolpropane, in the backbone, so that they contain pendant hydroxyls which act as the sites of crosslinking. A number of different di-isocyanates are used commercially; typical examples are shown in Table 1.2.

The major use of polyurethanes is as rigid or flexible foams. The rigid polyurethanes have a variety of uses, including insulating material for filling cavity walls of houses. Flexible polyurethanes have been widely used in soft furnishing. They suffer from the considerable disadvantage that they burn easily with a very smokey and toxic flame, which has led to severe problems in accidental domestic fires. The use of these materials for this purpose is now declining.

1.4.12 Poly(ether ether ketone)

Poly(ether ether ketone), known as **PEEK**, is a crystalline high-temperature thermoplastic. It is produced by a step (condensation)

Table 1.2 Commercially important di-isocyanates.

Name	Abbreviation	Structure
Tolylene di-isocyanate	TDI	(structure: methyl-substituted benzene ring with two NCO groups)
Hexamethylene di-isocyanate	HMDI	$OCN(CH_2)_6NCO$
Diphenylmethane 4,4'-di-isocyanate	MDI	$OCN\!-\!\langle\rangle\!-\!CH_2\!-\!\langle\rangle\!-\!NCO$

process (see Chapter 2) in which 4,4-difluorobenzophenone is reacted with the anion of hydroquinone. This reaction is carried out in a high boiling point polar solvent, such as *N*-cyclohexyl-2-pyrrolidone (1.13).

(1.13)

An indication of the thermal stability of PEEK is given by the fact that it can be used at temperatures of up to 250 °C without loss of physical properties. Its melting temperature is 334 °C. The degree of crystallinity can be reduced by copolymerisation with monomers containing diphenyl units, thus reducing the degree of regularity in the linear array of linked phenyl rings.

Industrial applications of PEEK take advantage of the outstanding thermal properties, and include use in high-performance automotive and aerospace components, and in printed circuit boards.

1.4.13 Silicones

Silicones are polymers with alternating silicon and oxygen atoms in the backbone in which the silicon atoms are also joined to organic groups. Typical structures found are (1.4) and (1.5).

The possibility of all kinds of morphology from linear, low relative molar mass molecules to three-dimensional networks means that silicones are found in a range of physical forms, from fluids through to insoluble solids.

The name silicone was originally coined because of the resemblance of the empirical formulae of silicones, *e.g.* R_2SiO, with ketones, R_2CO. However, because of the different nature of the silicon and carbon atoms, in particular that silicon is too large to form stable double bonds, this resemblance is merely superficial.

Silicones are relatively expensive, but they do have excellent thermal stability and water repellency. Hence they tend to find application where these properties are particularly necessary.

1.4.14 Naturally Occurring Polymers

There are a number of naturally occurring polymers which find technical application, including cellulose and its derivatives, starch, and rubber. In addition, a number of important biological materials, most notably the proteins, are made up of macromolecules. These will be considered briefly in the sections which follow.

1.4.15 Cellulose

This is a very widely available polymer, since it is the main component of the cell walls of all plants. It is a carbohydrate of molecular formula $(C_6H_{10}O_5)_n$, where n runs to thousands. The cellulose 'monomer' is D-glucose, and the cellulose molecules are built up from this substance, effectively by condensation and removal of the elements of water.

D-Glucose itself is highly soluble in water, but cellulose is not. This is essentially a kinetic phenomenon; the hydroxy-groups in cellulose would, in principle, readily form hydrogen bonds with water molecules, and hence the cellulose macromolecule would be carried off into aqueous solution. But these hydroxy-groups interact with neighbouring cellulose molecules, making it impossible for water molecules to penetrate, or solvate, the individual molecules. Cellulose will, however, dissolve in the somewhat strange medium aqueous ammoniacal cupric hydroxide, $Cu(NH_3)_4(OH)_2$.

Cellulose is a linear polymer. Despite this, it is not thermoplastic, essentially because of its extensive intermolecular hydrogen bonding which never allows the molecules to move sufficiently for the polymer to melt.

Cellulose may be solubilised by treatment with sodium hydroxide and carbon disulfide. It can be regenerated by acidification of the solution. This is the basis of the production of regenerated cellulose fibre, so-called 'viscose rayon', which is a major textile fibre.

The technique is also used for the production of continuous cellulose-derived film, so-called 'cellophane' (from 'cellulose' and 'diaphane', the latter being French for transparent).

Cellulose is also commercially modified by acetylation to produce a material suitable for X-ray and cine film. Commercially cellulose ethers are also prepared, such as methylcellulose. This material is water-soluble and gives a highly viscous solution at very low concentrations. Hence it is widely used as a thickener in latex paints and adhesives, in cosmetics and for coating pharmaceutical tablets.

1.4.16 Starch

Starch is a widely distributed material which occurs in roots, seeds, and fruits of plants. For commercial use, corn is the principal source, though wheat and potatoes are also used. Starch is extracted by grinding with water, filtering, centrifuging, and drying, a process which yields starch in a granular form.

Starch consists of two components, amylose and amylopectin. The ratio of these varies with the source of the starch, with amylopectin usually predominating and representing some 70–85% of the total mass of the starch.

Amylopectin is the polymeric component of starch and consists mainly of glucose units joined at the 1,4-positions. Relative molar mass tends to be very high, *e.g.* between 7 and 70 million. A variety of modified starches are used commercially which are produced by derivatisation to give materials such as ethanoates (acetates), phosphates, and hydroxyalkyl ethers. Modified and unmodified starches are used in approximately equal tonnages, mainly in paper-making, paper coatings, paper adhesives, textile sizes, and food thickeners.

1.4.17 Natural Rubber

Rubber is obtained from the juice of various tropical trees, mainly the tree *Hevea brasiliensis*. The juice is a latex consisting of a dispersion of polymer phase at a concentration of about 35% by mass, together with traces of proteins, sterols, fats, and salts. The rubber is obtained either by coagulation of the latex with acid, either ethanoic or methanoic, or by evaporation in air or over a flame. The material that results from this process is a crumbly, cheeselike substance, sometimes called raw rubber or caoutchouc. In order to develop the mechanical properties that are considered characteristic of rubber, *i.e.* so-called rubberlike elasticity, this raw rubber needs further processing, and in particular lightly crosslinking. This is achieved in the process known as vulcanisation, as will be discussed later.

The polymer in natural rubber consists almost entirely of *cis*-poly(isoprene) (1.6). The molecules are linear, with relative molar mass typically lying between 300 000 and 500 000. The macromolecular nature of rubber was established mainly by Staudinger in 1922, when he hydrogenated the material and obtained a product that retained its colloidal character, rather than yielding fragments of low relative molar mass.

$$\begin{array}{cc} CH_3 & H \\ \diagdown \;\; \diagup \\ C{=}C \\ \diagup \;\;\; \diagdown \\ -CH_2 & CH_2{-} \end{array}$$

(1.6)

Vulcanisation is the term used for the process in which the rubber molecules are lightly crosslinked in order to reduce plasticity and develop elasticity. It was originally applied to the use of sulfur for this purpose, but is now used for any similar process of cross-linking. Sulfur, though, remains the substance most widely used for this purpose.

Sulfur reacts very slowly with rubber, and so is compounded with rubber in the presence of accelerators and activators. Typical accelerators are thiazoles and a typical activator is a mixture of zinc oxide and a fatty acid. The chemistry of the vulcanisation reactions is complicated, but generates a three-dimensional network in which rubber molecules are connected by short chains of sulfur atoms, with an average of about five atoms in each chain.

A much more heavily crosslinked material can be obtained by increasing the amount of sulfur in the mixture, so that it represents about a third of the mass of the product. Heating such a mixture of raw rubber and sulfur at 150 °C until reaction is complete gives a hard, thermoset material that is not at all elastic. This material is called ebonite and is used to make car battery cases.

1.4.18 Proteins

The proteins are a group of macromolecular substances of great importance in biochemistry. Their very name provides testimony to this – it was coined by Mulder in 1838 from the Greek word 'proteios', meaning 'of first importance'. They appear in all cells, both animal and plant, and are involved in all cell functions.

Proteins are linear polyamides formed from α-amino acids. An α-amino acid is one in which carboxylic acid and the amino group reside on the same carbon atom (1.7).

In nature, there are 20 amino acids available for incorporation into the protein chain. They are arranged in a specific and characteristic sequence along the molecule. This sequence is generally referred to as the 'primary structure' of the protein. Also part of the primary structure is the relative molar mass of the macromolecule.

$$R-\underset{\underset{NH_2}{|}}{CH}-COOH$$

(1.7)

As a result of the particular amino acid sequence, the protein molecule adopts a characteristic arrangement, such as symmetrical coils or orderly foldings. This is known as the 'secondary structure'. Such individual coils or folded structures bring about a longer range stable arrangement called the 'tertiary structure'. Lastly, several protein molecules may join together to form a complex, this final grouping being known as the 'quaternary structure'.

Proteins themselves consist of large molecules with several hundred amino acids in the primary structure. Smaller units, known as polypeptides, may be formed during physiological processes. There is no clear distinction between a protein and a polypeptide; about 200 amino acid units in the primary structure is usually taken to be the minimum for a substance to be considered a protein.

The diversity in primary, secondary, tertiary, and quaternary structures of proteins means that few generalisations can be made concerning their chemical properties. Some fulfil structural roles, such as the collagens (found in bone) and keratin (found in claws and beaks), and are insoluble in all solvents. Others, such as albumins or globulins of plasma, are very soluble in water. Still others, which form part of membranes of cells, are partly hydrophilic ('water-loving', hence water-soluble) and partly lipophilic ('lipid-loving', hence fat-soluble).

Because proteins are not composed of identical repeating units, but of different amino acids, they do not fall within the formal definition of polymers given at the start of this chapter. They are nevertheless macromolecular and techniques developed for the study of true polymers have been applied to them with success. However, for the most part they are outside the scope of this book and accordingly will receive very little attention in the chapters that follow.

1.4.19   Poly(3-hydroxybutyrate)

Poly(3-hydroxybutyrate) (1.8) is a bacterial polyester that behaves as an acceptable thermoplastic, yet can be produced from renewable agricultural feedstocks and is biodegradable. It is typically produced not in the pure state, but formed alongside minor amounts of poly(3-hydroxyvalerate). The ratio of these two polymers in a given sample is determined by the ratio of glucose and propionic acid in the medium in which the bacteria live and carry out their metabolic processes.

$$\left(O-\underset{\underset{H}{|}}{\overset{\overset{CH_3}{|}}{C}}-CH_2-\overset{\overset{O}{||}}{C} \right)_n$$

(1.8)

The carbon atom which carries the methyl group is chiral, but biosynthesis is stereoselective, and gives rise to a natural polymer with the *R* configuration. The polymer is a partially crystalline thermoplastic which melts at about 80 °C.

Poly(3-hydroxybutyrate) has attracted interest as an environmentally degradable thermoplastic that can be used in packaging, agriculture and medicine. It undergoes enzymic degradation quite readily, becoming less crystalline and rapidly decreasing in molar mass. The rate of degradation depends on a number of factors, such as moisture level, nutrient supply, temperature, and pH. Further details are given in Chapter 8.

2 Polymerisation Reactions

As outlined in Chapter 1, polymerisation reactions can be classified as either condensation or addition processes, the basis of the classification suggested by W. H. Carothers in 1929. More useful, however, is the classification based on reaction kinetics, in which polymerisation reactions are divided into step and chain processes. These latter categories approximate to Carothers' condensation and addition reactions but are not completely synonymous with them.

The study of reaction mechanisms can be a subtle business but in fact the mechanistic basis of classification into step and chain processes arises from major differences in the two types of process. There is no doubt about the nature of the reaction in almost all cases as can be seen by considering the distinguishing features of the two mechanisms which are summarised below.

Chain polymerisation: Monomer concentration decreases steadily with time. High molar mass polymer is formed at once and the molar mass of such early molecules hardly changes at all as reaction proceeds. Long reaction times give higher yields but do not affect molar mass. The reaction mixture contains only monomer, high molar mass polymer, and a low concentration of growing chains.

Step polymerisation: Monomer concentration drops rapidly to zero early in the reaction. Polymer molar mass rises steadily during reaction. Long reaction times increase molar mass and are essential to obtain very high molar masses. At all stages of the reaction every possible molecular species from dimers to polymers of large degrees of polymerisation are present in a calculable distribution.

The Chemistry of Polymers, 5th Edition
By John W. Nicholson
© John W. Nicholson 2017
Published by the Royal Society of Chemistry, www.rsc.org

2.1 Chain Polymerisation

The polymerisation reactions that occur by the chain mechanism are typically those involving unsaturated monomers. The characteristic reaction begins with the chemical generation of reactive centres on selected monomer molecules. These reactive centres are typically free radicals, but may be anions or cations, and they react readily with other monomers without extinguishing the active centre. In this way any given active centre becomes responsible for the reaction of large numbers of monomer molecules which add to the growing polymer, thereby increasing its molar mass. The reactivity of the species involved is high, hence so is the rate constant of reaction in a typical polymerisation, which leads to the rapid formation of high molar mass polymers right from the beginning of the reaction. A consequence of this is that almost no species intermediate between monomer and polymer are found in such reacting systems.

Chain reactions do not continue indefinitely, but in the nature of the reactivity of the free radical or ionic centre they are likely to react readily in ways that will destroy the reactivity. For example, in radical polymerisations two growing molecules may combine to extinguish both radical centres with formation of a chemical bond. Alternatively they may react in a disproportionation reaction to generate end groups in two molecules, one of which is unsaturated. Lastly, active centres may find other molecules to react with, such as solvent or impurity, and in this way the active centre is destroyed and the polymer molecule ceases to grow.

Chain polymerisation typically consists of these three phases, namely initiation, propagation, and termination. Because the free-radical route to chain polymerisation is the most important, both in terms of versatility and in terms of tonnage of commercial polymer produced annually, this is the mechanism that will be considered first and in the most detail.

2.1.1 Initiation

The monomers used in chain polymerisations are unsaturated, sometimes referred to as vinyl monomers. In order to carry out such polymerisations a small trace of an initiator material is required. These substances readily fragment into free radicals either when heated or when irradiated with electromagnetic radiation from around or just beyond the blue end of the spectrum. The two most commonly used free radical initiators for these reactions are benzoyl peroxide and azobisisobutyronitrile (usually abbreviated to AIBN). They react as indicated in Reactions (2.1) and (2.2).

$$\text{Benzoyl peroxide} \quad C_6H_5\text{-}CO_2\text{-}O_2C\text{-}C_6H_5 \longrightarrow 2\ C_6H_5\text{-}CO_2\text{•} \tag{2.1}$$

Benzoyl peroxide

$$(CH_3)_2\overset{|}{\underset{CN}{C}}N{=}N\overset{|}{\underset{CN}{C}}(CH_3)_2 \longrightarrow 2\ (CH_3)_2\overset{|}{\underset{CN}{C}}\text{•} \ +\ N_2 \tag{2.2}$$

In addition to heat and light, generation of free radicals can be accomplished by using γ-rays, X-rays or through electrochemical means. In general, however, these methods do not tend to be so widely used as those involving benzoyl peroxide or AIBN as initiators.

Once produced the free radical reacts rapidly with a molecule of monomer to yield a new species that is still a free radical as shown in Reaction (2.3).

$$R\text{•} \ +\ CH_2{=}CHX \longrightarrow RCH_2\overset{\overset{\displaystyle H}{|}}{\underset{\underset{\displaystyle X}{|}}{C}}\text{•} \tag{2.3}$$

The efficiency of the intitiator is a measure of the extent to which the number of radicals formed reflects the number of polymer chains formed. Typical initiator efficiencies for vinyl polymerisations lie between 0.6 and 1.0. Clearly the efficiency cannot exceed 1.0 but it may fall below this figure for a number of reasons, the most important being the tendency of the newly generated free radicals to recombine before they have time to move apart. This phenomenon is called the 'cage effect'.

2.1.2 Propagation

This is the name given to the series of reactions in which the free radical unit at the end of the growing polymer molecule reacts with monomer to increase still further the length of the polymer chain. They may be represented as shown in Reaction (2.4).

$$R(CH_2CHX)_n-CH_2\overset{\overset{\displaystyle H}{|}}{\underset{\underset{\displaystyle X}{|}}{C}}\bullet \quad + \quad CH_2{=}CHX \quad \longrightarrow$$

$$(2.4)$$

$$R(CH_2CHX)_{(n+1)}-CH_2\overset{\overset{\displaystyle H}{|}}{\underset{\underset{\displaystyle X}{|}}{C}}\bullet$$

2.1.3 Termination

Polymerisation does not continue until all of the monomer is used up because the free radicals involved are so reactive that they find a variety of ways of losing their radical activity. The two methods of termination in radical polymerisations are combination and disproportionation. The first of these occurs when two radical species react together to form a single bond and one reaction product as in Reaction (2.5).

$$-CH_2\overset{\overset{\displaystyle H}{|}}{\underset{\underset{\displaystyle X}{|}}{C}}\bullet \;+\; \bullet\overset{\overset{\displaystyle H}{|}}{\underset{\underset{\displaystyle X}{|}}{C}}CH_2- \;\longrightarrow\; -CH_2\overset{\overset{\displaystyle H}{|}}{\underset{\underset{\displaystyle X}{|}}{C}}-\overset{\overset{\displaystyle H}{|}}{\underset{\underset{\displaystyle X}{|}}{C}}CH_2- \quad (2.5)$$

Alternatively, two radicals can interact via hydrogen abstraction, leading to the formation of two reaction products, one of which is saturated and one of which is unsaturated. This is known as disproportionation (Reaction (2.6)).

$$-CH_2\overset{\overset{\displaystyle H}{|}}{\underset{\underset{\displaystyle X}{|}}{C}}\bullet \;+\; \bullet\overset{\overset{\displaystyle H}{|}}{\underset{\underset{\displaystyle X}{|}}{C}}CH_2- \;\longrightarrow\; -CH_2\overset{\overset{\displaystyle H}{|}}{\underset{\underset{\displaystyle X}{|}}{C}}-H \;+\; \overset{\overset{\displaystyle H}{|}}{\underset{\underset{\displaystyle X}{|}}{C}}{=}CH- \quad (2.6)$$

Both termination mechanisms have been shown to occur experimentally, the method being to examine the polymer molecules formed for fragments of initiator. In such a way polystyrene has been found to terminate mainly by combination and poly(methyl methacrylate) entirely by disproportionation at temperatures above 60 °C.

2.1.4 Other Reactions

Chain polymerisation necessarily involves the three steps of initiation, propagation, and termination, but the reactivity of the free radicals is such that other processes can also occur during polymerisation. The major one is known as chain transfer and occurs when the reactivity of the free radical is transferred to another species which in principle is capable of continuing the chain reaction. This chain transfer reaction thus stops the polymer molecule from growing further without at the same time quenching the radical centre.

Typical chain transfer reactions involve the abstraction of an atom from a neutral saturated molecule, which may be solvent or a chain transfer agent added to the polymerisation mixture specifically to control the final size and distribution of molar masses in the polymer product. The chain transfer reaction may be represented as in Reaction (2.7).

$$
\overset{\overset{\displaystyle H}{|}}{-CH_2\underset{\underset{\displaystyle X}{|}}{C}}\cdot \; + \; CCl_4 \longrightarrow -CH_2\overset{\overset{\displaystyle H}{|}}{\underset{\underset{\displaystyle X}{|}}{C}}-Cl \; + \; \cdot CCl_3 \qquad (2.7)
$$

The newly generated free radical is then free to react with a molecule of monomer and thus lead to the establishment of a new polymer molecule.

Other chain transfer processes may occur. For example, the radical may abstract an atom from along the backbone of a previously formed polymer molecule, and thus initiate the growth of a branch to the main chain. There can also be chain transfer to monomer, which in the nature of the polymerisation process must be a relatively rare phenomenon. However, it can occur infrequently and give rise to a restriction in the size of the polymer molecules without ceasing the overall radical chain reaction.

The reactions mentioned so far all take place with the generation of a free radical of high reactivity which is capable of sustaining the chain reaction. However, other molecules exist which form free radicals of such high stability that they effectively stop the chain process. These molecules are called retarders or inhibitors; the difference is one of degree, retarders merely slowing down the polymerisation reaction while inhibitors stop it completely. In practice vinyl monomers such as styrene and methyl methacrylate are stored with a trace of inhibitor in them to prevent any uncontrolled polymerisation before use. Prior to polymerisation these liquids must be freed from this inhibitor, often by aqueous extraction and/or distillation.

2.2 Arrangement of Monomer Units

In chain reactions the addition of vinyl monomers to a free radical can occur in one of two ways (Reaction (2.8)).

$$R\cdot \quad + \quad CH_2{=}CHX \quad \Bigg\langle \quad \begin{array}{l} R{-}CH_2{-}\underset{\underset{X}{|}}{CH}\cdot \\[2em] R{-}\underset{\underset{X}{|}}{CH}{-}CH_2\cdot \end{array} \tag{2.8}$$

The first of these leads to the formation of a completely regular arrangement, described as a *head-to-tail* configuration. In it the substituents occur on alternate carbon atoms along the polymer molecule. The second arrangement gives an irregular arrangement containing either *head-to-head* or *tail-to-tail* configurations. Hence substituents occur at irregular intervals along the backbone. In fact these latter configurations are generally fairly rare, though they do appear occasionally in polymer molecules. Many factors contribute to this rarity, including steric effects which favour the head-to-tail arrangement, so that vinyl polymers are most commonly encountered with this particular structure.

2.3   Kinetics of Chain Polymerisation

Mathematical expressions can be derived to describe the kinetics of chain processes building on the ideas already discussed that such reactions take place in the well defined stages of initiation, propagation, and termination.

Initiation occurs as a molecule of intiator decomposes to free radicals (Reaction (2.9)), where k_d is the rate constant for the reaction. These newly generated free radicals can then interact with monomer molecules preserving the free radical centre (Reaction (2.10)). These two processes may be considered to form part of the initiation stage, the first of them being rate determining.

Following initiation comes the series of propagation steps, which can be generalised as in Reaction (2.11). A single rate constant, k_p, is assumed to apply to these steps since radical reactivity is effectively independent of the size of the growing polymer molecule.

$$I \quad \xrightarrow{\ k_d\ } \quad 2R\cdot \tag{2.9}$$

$$R\cdot \quad + \quad M \quad \xrightarrow{\ k_a\ } \quad RM\cdot \tag{2.10}$$

$$M\cdot_n \quad + \quad M \quad \xrightarrow{\ k_p\ } \quad M\cdot_{(n+1)} \tag{2.11}$$

Termination may be the result of either combination (Reaction (2.12)) or disproportionation (Reaction (2.13)). However, it is rarely necessary to distinguish between these two termination mechanisms, and so the rate constants are generally combined into a single rate constant, k_t.

$$M\bullet_n \quad + \quad M\bullet_m \xrightarrow{\;k_{tc}\;} M_{(m+n)} \qquad (2.12)$$

$$M\bullet_n \quad + \quad M\bullet_m \xrightarrow{\;k_{td}\;} M_n \quad + \quad M_m \qquad (2.13)$$

The rates of the three steps may be written in terms of the concentrations of the chemical species involved and these rate constants. The rate of initiation then is given by:

$$R_i = \frac{d[M^\bullet]}{dt} = 2fk_d[I]$$

where square brackets mean 'concentration of'. The factor f is introduced because of the cage effect and represents the fraction of initiator fragments that bring about polymerisation.

The rate of termination is given by:

$$R_t = -\left(\frac{d[M^\bullet]}{dt}\right)_t = 2k_t[M^\bullet]^2$$

Early on in the overall polymerisation the rates of initiation and termination become equal, resulting in a steady-state concentration of free radicals. Here $R_i = R_t$. From this the two equations can be combined to solve for $[M^\bullet]$:

$$[M^\bullet] = \left(\frac{fk_d[I]}{k_t}\right)^{1/2}$$

Since propagation is the stage that involves the major consumption of monomer (otherwise polymer would not form), the rate of monomer loss can be expressed in terms of propagation only:

$$R_p = -\frac{d[M]}{dt} = k_p[M][M^\bullet]$$

which with substitution becomes:

$$R_p = k_p \left(\frac{fk_d[I]}{k_t}\right)^{1/2} [M]$$

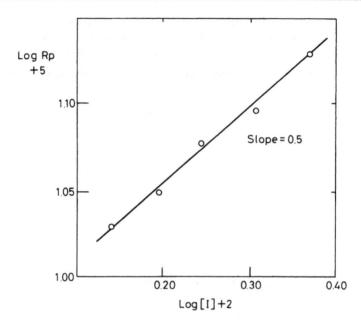

Figure 2.1   Example of the relationship between rate of reaction, R_p, and initiator concentration, [I] for the iso-quinoline-bromine complex in methyl methacrylate.
(J. W. Nicholson, unpublished results, 1977.)

This mathematical treatment shows that in the early stages of polymerisation the rate of reaction should be proportional to the square root of the initiator concentration, assuming f is independent of monomer concentration. This assumption is acceptable for high initiator efficiencies, but with very low efficiencies, f may become proportional to [M], making the rate proportional to $[M]^{3/2}$.

The finding that the rates of chain polymerisations are proportional to the square root of the initiator concentration is well established for a large number of polymerisation reactions. An example is shown in Figure 2.1, which also illustrates the method by which such initiator exponents are determined, *i.e.* by a plot of $\log R_p$ *vs.* $\log[I]$.

2.4   Autoacceleration

For the major duration of a chain polymerisation the reaction is first-order in monomer concentration. However, at high conversions of monomer to polymer using either undiluted monomer or concentrated solutions there is a significant deviation from first-order kinetics. Under such circumstances the rate of reaction (and also molar mass of polymer) increases considerably. This so-called *auto-acceleration* is sometimes referred to as the Trommsdorff–Norrish effect, after two of the pioneers in the study of polymerisation kinetics, who first noticed its occurrence.

The explanation for autoacceleration is as follows. As polymerisation proceeds there is an increase in the viscosity of the reaction mixture which reduces the mobility of the reacting species. Growing polymer molecules are more affected by this than either the molecules of monomer or the fragments arising from decomposition of the initiator. Hence termination reactions slow down and eventually stop, while initiation and propagation reactions still continue. Such a decrease in the rate of the termination steps thus leads to the observed increase in the overall rate of polymerisation.

2.5 Practical Methods of Chain Polymerisation

Chain reactions are used to prepare a variety of high molar mass polymers of commericial importance and in practice may take one of four forms, namely bulk, solution, suspension, and emulsion methods. These four methods are described in the sections that follow, together with the 'loop' modification which has become of commercial importance recently in producing latexes by emulsion polymerisation for the paint industry.

2.5.1 Bulk Polymerisation

At first sight this technique appears to be the method of choice for producing polymers by chain processes. Since the starting material consists mainly of pure polymer, with only traces of initiator and possible chain transfer agent present (the latter to modify the molar mass and polydispensity), this appears to be the ideal technique for the preparation of high molar mass polymer. However, there are problems with this system. For example viscosity increases as polymerisation occurs and this results in difficulties in handling the product. Also, since chain reactions are generally exothermic, and the increasing viscosity inhibits dissipation of heat, there can be localised overheating leading to charring and possible degradation of the product.

For these reasons, despite the apparent advantages and also despite the fact that bulk polymerisation is so often the method of choice for the laboratory preparation of vinyl polymers, this technique is not widely used in industry. Only three polymers are produced in this way, namely poly(ethylene), poly(styrene), and poly(methyl methacrylate).

Of these polymers, poly(ethylene) is produced from a gaseous monomer under pressure, either high or low, and thus some of the disadvantages mentioned above for bulk polymerisation hardly apply. They are certainly more severe for polymerisation of liquid monomers, such as styrene and methyl methacrylate. In the case of poly(styrene), bulk polymerisation is nonetheless used for the commercial production of the polymer. Ingenious engineering of the plant overcomes the problem of the exotherm: manufacture of the polymer takes place in discrete stages in different parts of the plant. The reaction is initiated in a tank which is heated to a temperature of 80 °C; styrene undergoes self-initiation on heating, so that no extra initiator is required for this step, which is allowed to continue until about 35% conversion to polymer. At this conversion the mixture still has a sufficiently low viscosity to enable fairly easy stirring and transport.

From this stage of 35% polymerisation the mixture is passed down a tower in an atmosphere of nitrogen; there is a thermal gradient throughout the tower from 100 °C at the top to 200 °C at the bottom. This gradient is maintained by a complicated arrangement of heaters and coolers which compensate for the exotherm that the poly(styrene) undergoes itself as increasing proportions of monomer are converted to polymer.

At the bottom of the tower, the high molar mass poly(styrene) is extruded, granulated, and cooled prior to packaging.

2.5.2   Solution Polymerisation

One way of overcoming some of the problems associated with bulk polymerisation is to dissolve the monomer in an appropriate solvent. In particular difficulties associated with the exotherm on polymerisation may be overcome since temperature can be more readily controlled than in the bulk technique. If the right solvent is chosen the product may form to give a solution suitable for casting or spinning. There are disadvantages with solution polymerisation, though. Firstly, reaction temperature is limited by the boiling point of the solvent used which in turn restricts the rate of reaction that may be achieved. Secondly, it is difficult to free the product of the last traces of the solvent. Finally, selection of a completely inert solvent cannot actually be done, which means that there is almost always chain transfer to the solvent and hence a restriction on the molar mass of the product. This last point is particularly important and is the one that is primarily responsible for the rarity of solution techniques in the manufacture of commercially important polymers.

2.5.3 Suspension Polymerisation

This technique is very similar to solution polymerisation except that the monomer is suspended rather than dissolved in an inert liquid, often water. Heat transfer and reduction in viscosity are comparable with those of solution polymerisations, though mechanical agitation and the presence of suspending agents are necessary to maintain the monomer in suspension. Effectively the technique works because there are a large number of microdroplets undergoing bulk polymerisation.

There are disadvantages with suspension polymerisation. In particular, for polymers that are very soluble in their monomer, stirring has to be extremely vigorous, otherwise the partially reacted droplets undergo agglomeration. Also, tacky polymers (such as synthetic elastomers) are very prone to undergo agglomeration, so that suspension polymerisation cannot be used for these polymers.

Despite these disadvantages the suspension technique is used industrially, poly(vinyl chloride) being produced by this method. For this process vinyl chloride monomer is suspended in demineralised water to which gelatin has been added as suspending agent and caproyl peroxide added as initiator. Polymerisation takes place under an atmosphere of nitrogen at 50 °C, taking 12 hours to reach 85–90% completion, at which point reaction is stopped. Pure PVC is obtained from the suspension by centrifugation.

2.5.4 Emulsion Polymerisation

Emulsion polymerisation represents the next stage in development from the suspension technique and is a versatile and widely used method of polymerisation. In this technique droplets of monomer are dispersed in water with the aid of an emulsifying agent, usually a synthetic detergent. The detergent forms small micelles 10–100 μm in size, which is much smaller than the droplets that can be formed by mechanical agitation in suspension polymerisation. These micelles contain a small quantity of monomer, the rest of the monomer being suspended in the water without the aid of any surfactant.

Emulsion polymerisation is initiated using a water-soluble initiator, such as potassium persulfate. This forms free radicals in solution which may initiate some growing chains in solution. These radicals or growing chains pass to the micelles and diffuse into them, which causes the bulk of the polymerisation to occur in these stabilised droplets.

As emulsion polymerisation proceeds, like the suspension technique but unlike either the bulk or the solution techniques, there is almost no increase in viscosity. The resulting dispersed polymer is not a true emulsion any more, but instead has become a latex. The particles of the latex do not interact with the water; hence viscosity is not found to change significantly up to about 60% solids content.

Emulsion polymerisation is used in the commercial production of synthetic diene elastomers and also to produce commercial latexes of the type used in paints; these paints are known incorrectly as 'emulsion' paints and are used extensively in D.I.Y. For synthetic elastomers, the latex is 'broken' by changing the pH of the water, which causes rapid agglomeration of the particles and hence what is effectively precipitation. The solid product is washed on a rotary filter to remove the detergent then dried in air prior to being packed and sold.

2.5.5 The Loop Process

This is a modification of emulsion polymerisation which has recently been developed for the manufacture of commercially important latexes for 'emulsion' paints. In this process instead of producing the polymer in batches in a tank polymer is produced continuously in a reactor that consists of a continuous tube coiled to a convenient shape.

The loop apparatus was invented in the early 1970s and was designed as an aid for kinetic studies. It works on the principle that by initiating reaction at one point in the coiled reactor, and continuously circulating the reacting emulsion, the system eventually settles down to give a steady-state concentration of free radicals at any point in the loop. Raw materials are added in separate streams, the monomer in one, together with one of the initiator components, and surfactants, buffers, and the other initiator component in the other. Two-component initiators are used to give redox initiation as the route to free radicals. The product is drawn off at the same rate as reactants are added, and since its properties do not vary with time, the rate at which new polymer particles form must equal the rate at which the old ones are lost as the product is drawn off.

The major advantage of the loop process is that the equipment required to run it is small, lightweight, and hence of low cost. It needs minimal supervision yet can produce significant volumes of latex over the course of a year, and so is becoming increasingly important in industry.

2.6 Other Chain Polymerisation Mechanisms

Although the most important chain reactions are those involving free radicals, they are not the only ones that are possible. The reactive centre at the growing end of a polymer molecule may alternatively be ionic in character or involve co-ordination to metal complexes.

Ionic polymerisation is subdivided into cationic and anionic mechanisms depending on the charge developed in the growing polymer molecule. Typical catalysts for the former, the cationic polymerisation process, are Lewis acids such as $AlCl_3$ or BF_3, which often require a co-catalyst, usually a Lewis base, in order to bring about polymerisation.

It is characteristic of cationic polymerisation that high rates of reaction can be obtained at low temperatures. For example isobutylene [2-methylpropene, $(CH_3)_2 C=CH_2$] can be polymerised by BF_3 to yield polymers of molar mass well in excess of one million in just a few seconds. The first step in the reaction is between BF_3 and the co-catalyst, usually water. The resulting complex readily protonates an isobutylene molecule to yield the carbenium ion $C(CH_3)_3^+$, which is the species that starts the polymerisation proper. For energetic reasons, the process can only yield head-to-tail polymer.

Termination in such a system occurs in what is essentially a reversal of the initiation reaction, *i.e.* protonation of the $(BF_3OH)^-$ species to yield a polymer containing an unsaturated end group (2.1).

$$CH_3[(CH_3)_2(CH_2)]_n\overset{\overset{\displaystyle CH_3}{|}}{C}=CH_2$$

(2.1)

Anionic polymerisation involves the development of a negative charge on the growing polymer molecule. This is achieved through the use of a catalyst that can readily form anions which themselves react with the vinyl monomer. An example is shown in Reaction

(2.14). Here the initiator is sodium and the reaction is carried out in liquid ammonia at $-75\ °C$.

$$NH_2^- \quad + \quad CH_2{=}CHX \quad \longrightarrow \quad H_2NCH_2CHX^- \qquad (2.14)$$

As is the case for cationic polymerisation, anionic polymerisation can terminate by only one mechanism, that is by proton transfer to give a terminally unsaturated polymer. However, proton transfer to initiator is rare – in the example just quoted, it would involve the formation of the unstable species NaH containing hydride ions. Instead proton transfer has to occur to some kind of impurity which is capable for forming a more stable product. This leads to the interesting situation that where that monomer has been rigorously purified, termination cannot occur. Instead reaction continues until all of the monomer has been consumed but leaves the anionic centre intact. Addition of extra monomer causes further polymerisation to take place. The potentially reactive materials that result from anionic initiation are known as *'living' polymers*.

Finally, chain polymerisation can occur *via* coordination, as is the case for polymerisation involving Ziegler–Natta catalysts. These catalysts are complexes formed between main-group metal alkyls and transition metal salts. Typical components are shown in Table 2.1.

The components of the Ziegler–Natta catalyst unite in such a way that there is a vacant co-ordination site on the transition metal to which a molecule of monomer can bond, a second molecule of monomer then attaches itself to another vacant coordination site, from which it is able to react with the first molecule of coordinated monomer. This causes the second coordination site to become vacant once again, thereby allowing a further molecule of monomer to enter and react with the now growing polymer chain. Because of the spatial requirements of the transition metal co-ordination compound, this kind of polymerisation tends to continue along highly stereospecific lines, producing polymer in which all of the monomer units have identical stereochemistries, *i.e.* so-called *isotactic* polymer.

Table 2.1 Components of typical Ziegler–Natta catalysts.

Main group metal alkyls	Transition metal salts
$(CH_3CH_2)_3Al$	$TiCl_4$
$(CH_3CH_2)_2AlGl$	VCl_3
C_4H_9Li	
C_4H_9MgI	

The monomers that can be polymerised using these catalysts are restricted to those without polar functional groups, *e.g.* ethylene, propylene, but-1-ene, butadiene, and styrene. Chain transfer agents are required to prevent build-up of excessive molar masses. The commonest such transfer agent is hydrogen, H_2, which acts by donating one atom to the end of the polymer molecule, and the other to the active co-ordination site of the catalyst. This is then available to form the end-group of the next polymer chain which grows on that particular co-ordination site.

A wide variety of Ziegler–Natta catalysts have been studied in the laboratory, notably compounds of zirconium and hafnium, but commercial catalysts tend to be restricted to $TiCl_4$. This is used supported on magnesium chloride as the main catalyst component, with triethylaluminium, $(CH_3CH_2)_3Al$, as co-catalyst and diphenyl-dimethoxysilane, $(C_6H_5)_2Si(OCH_3)_2$, as external modifier. The role of the co-catalyst is to activate a co-ordination site on the $TiCl_4$ by reacting to form $(CH_3CH_2)_2AlCl$, leaving an ethyl group attached to the titanium atom. The external modifier has the role of ensuring that no other coordination sites around the titanium become activated and thus to prevent competing polymerisations.

2.7 Recent Developments in "Living" Polymerisation

2.7.1 RAFT Polymerisation

An alternative "living polymerisation" process is one involving free radicals and known as Reversible Addition Fragmentation (Chain) Transfer, or RAFT for short. It was first reported in 1998, and is now an established technique for synthesising polymers of varying structure and molar mass. The resulting polymers can also be prepared with a variety of reactive moieties at the chain ends.

In conventional free radical polymerisation, chain transfer agents, such as thiols, are routinely used to control molar mass. The result of reaction between a growing polymer chain that contains a radical centre and a thiol is a completed polymer chain that contains no unpaired electrons, *i.e.* one that can be considered "dead". However, if the chain transfer reaction were reversible, a radical centre capable of promoting further growth in molar mass could be re-established. This is the basis of RAFT polymerisation.

The essential steps of RAFT polymerisation are as follows:

$$P_n^\bullet + S{=}C{-}S{-}P_m \rightleftharpoons P_n{-}S{-}C{-}S{-}P_m \rightleftharpoons P_n{-}S{-}C{=}S + P_m^\bullet \qquad (2.15)$$
$$\quad\quad\;\; | \quad\quad\quad\quad\quad | \quad\quad\quad\quad\quad |$$
$$\quad\quad\;\; Z \quad\quad\quad\quad\quad Z \quad\quad\quad\quad\quad Z$$

(I)   (II)   (III)   (IV)   (V)

The key component of such polymerisation is the reagent that acts as both chain transfer agent and initiator, generally termed a RAFT agent. This compound is typically a thiocarbonyl compound or dithioester. In the course of the reaction, it reacts with a propagating radical to form an adduct radical. This adduct radical then fragments to form a polymeric thiocarbonylthio species and a different propagating radical.

All of these reactions are reversible, so that eventually equilibrium is established between the two possible propagating radicals (I and V) and the two dominant polymer species (II and IV). By using an excess of RAFT agent relative to initiator, the equilibrium can be made to favour the thiocarbonylthio terminated species. This means that the polymer chains are dormant.

The reaction can be expressed in simplified form by the reaction:

$$\text{Initiator} + \text{Monomer} + R{-}S{-}C({=}S){-}Z \rightarrow R{-}\text{Polymer}{-}S{-}C({=}S){-}Z \qquad (2.16)$$

The effectiveness of RAFT polymerisation is determined substantially by the choice of RAFT agent. How well a compound can act as a RAFT agent is determined by the nature of the leaving group, R, and the stabilizing group, Z. For a successful RAFT polymerisation, the rate of transfer between the free radicals I and II and the intermediate III should be high.

Some of the molecules that are able to do this, and hence act as effective RAFT agents are shown in Table 2.2.

The temperature has to be chosen carefully to ensure that the rate of polymerisation is appropriate and that the free radicals are produced from the source at the correct rate. At this temperature, too, the

Table 2.2 Typical RAFT agents for polymerisation of methacrylates and methacrylic acid.

$PhC{=}S{\cdot}SC(CN)(CH_3)_2$
$C_{12}H_{25}SC{=}S{\cdot}SC(CN)(CH_3)_2$
$PhCH_2C{=}S{\cdot}SC(Ph)(CH_3)_2$

RAFT equilibrium needs to favour the active rather than the dormant state. The temperature at which all of these conditions are fulfilled varies for each monomer/solvent/RAFT agent combination, and must be identified experimentally prior to deploying the synthesis routinely for a particular system.

Once the appropriate conditions are established, RAFT polymerisation can be used with a wide range of monomers, including (meth)acrylates, acrylamides, acrylonitrile, styrene, butadiene, vinyl acetate and *N*-vinylpyrrolidone. Indeed, there are now commercially available RAFT agents for most types of monomer that will undergo free radical polymerisation. RAFT also has the advantage that it is capable of occurring in a variety of solvents (or none), including water.

RAFT is a controlled process, so can be used to prepare polymers of low polydispersity and also well defined molar mass. It can also be used to synthesise polymers with complicated structures, such as comb, star and brush morphologies, as well as linear copolymers with tightly controlled amounts of the respective monomers in each of the blocks.

Polymers produced by RAFT processes are typically discoloured, due to the presence of the RAFT agents. These thioesters are also subject to decomposition, which may give the polymer an unacceptable pungent odour. To an extent, these problems can be overcome by the use of appropriate purification steps.

2.7.2 Atom Transfer Radical Polymerisation

Atom transfer radical polymerisation, ATRP, is another example of living polymerisation that involves free radicals. Discovered in 1995, ATRP employs an alkyl halide as the initiator with a transition metal complex as catalyst. Initiation is brought about by reaction of the alkyl halide with the transition metal complex. This is a one-electron reaction in which the transition metal is oxidised and forms a compound with the halide, leaving the alkyl group as a free radical. The alkyl radical then goes on to initiate the polymerisation process. A typical scheme is as follows:

$$\text{R–Br} + \text{Cu(\textsc{i})Br/ligand} \rightleftharpoons \text{R}^{\bullet} + \text{Cu(\textsc{ii})Br}_2/\text{ligand} \qquad (2.17)$$

Then:

$$\text{R}^{\bullet} + n\text{M (monomer)} \rightarrow \text{R(M)}_n^{\bullet} \qquad (2.18)$$

And:

$$\text{R(M)}_n^{\bullet} + \text{Cu(\textsc{ii})Br}_2/\text{ligand} \rightleftharpoons \text{R(M)}_n\text{Br} + \text{Cu(\textsc{i})Br/ligand} \qquad (2.19)$$

Complexes of other transition metals have also been used for this sort of process, including complexes of iron, ruthenium, nickel and osmium. The catalyst has a critical role in determining the position of the equilibrium between active, propagating polymer and its inactive, dormant form. Equilibrium should lie on the side of the dormant form, which means that side reactions are suppressed. The resulting polymer then has low polydispersity.

The most widely used components of ATRP are copper(i) complexes, typically bromides, pyridine-based ligands and alkyl halides. The latter, which act as initiators for these processes, are also typically bromides, because they are more reactive than alkyl chlorides. A variety of functionalities can be present on the monomer, including amino, epoxy and vinyl groups. Monomers that can be polymerised by these methods include styrene, methyl methacrylate and acrylonitrile. ATRP reactions have been carried out in bulk monomer, and also with solvents, such as toluene, methanol, xylene, DMF or water.

2.7.3 Ring Opening Methathesis Polymerisation

ROMP is an addition polymerisation process that uses cyclic monomers, such as cyclopentene or norbornene to produce high molar mass linear polymers. The driving force for the polymerisation reaction is relief of ring strain, and it is for this reason that cyclic monomers are required. However, they must contain small rings in order to ensure there is sufficient ring strain to drive the process. A molecule such as cyclohexene has almost no ring strain, and hence almost completely fails to polymerise by this process. The essential reactions of a ROMP process are shown in Reaction (2.20), where norbornene is polymerised with a metal-carbene, with one initiation step and a series of propagation steps:

(2.20)

The catalyst employed in this scheme is a metal-carbene, but this compound may be generated *in situ*, as it is for example when $RuCl_3$ is used in conjunction with an alcohol to polymerise cyclic monomers.

Balancing the nature of the catalyst, monomer, and reaction conditions controls the structure of the resulting polymer. Homogeneous catalysts, such as those containing tungsten or molybdenum, have high rates of initiation and give good control of polydispersity and also of chain tacticity. However, only a limited number of monomers will react with these catalysts. Ruthenium-based catalysts, though having lower initiation rates, are capable of reacting with a wider range of monomers, though the slowness of the initiation typically leads to greater polydispersities.

ROMP reactions are strongly influenced by the choice of solvent, as this influences the nature and reactivity of the carbene species formed. This, in turn, influences the molar mass of the resulting polymer. Because of the nature of the starting compounds, ROMP is a process for producing unconventional polymers. Typical of the polymers produced are poly(dicyclopentadiene) and poly(cyclo-octene). Products retain a proportion of double bonds and this is useful as these bonds can by functionalised to give more complex polymers or hydrogenated to form more conventional polyalkenes.

2.8   Step Polymerisation

As explained earlier step polymerisations generally occur by condensation reactions between functionally substituted monomers. In order to obtain high molar mass products bifunctional reactants are used; monofunctional compounds are used to control the reaction while trifunctional species may be included in order to give branched or crosslinked polymers. A number of types of reaction may be involved, as described briefly in the following paragraphs.

Typically such reactions take place between reactive components, such as dibasic acids with diamines to give polyamides, or dibasic acids with diols to form polyesters. This reaction has an important modification in the case of nylon 6,6 [poly(hexamethylene adipamide)], where the initial product of reaction between hexamethylenediamine and adipic acid is a salt, which can be recrystallised readily in order to obtain the high-purity intermediate essential for conversion to high molar mass product. The condensation part of the reaction in this case is brought about by heating the intermediate salt.

In certain cases the organic dibasic acid is not sufficiently reactive for the purpose of polymerisation, and so it is replaced either with its anhydride or its acid chloride. For example polyamides (nylons) are often prepared by reaction of the acid chloride with the appropriate diamine. In the spectacular laboratory prepatation of nylon 6,6 this is done by interfacial polymerisation. Hexamethylenediamine is dissolved in water and adipyl chloride in a chlorinated solvent such as tetrachloromethane. The two liquids are added to the same beaker where they form two essentially immiscible layers. At the interface, however, there is limited miscibility and nylon 6,6 of good molar mass forms. It can then be continuously removed by pulling out the interface.

Ring-opening reactions may also be used in order to make polymers by the step polymerisation mechanism. One commercially important example of this is the manufacture of nylon 6, which uses caprolactam as the starting material and proceeds *via* the ring-opening reaction shown in Reaction (2.15).

2.9   Step Polymerisation with Polyfunctional Monomers

If monomers which have functionalities greater than 2 are used for step polymerisation the product that forms consists of an infinitely large three-dimensional network and the polymerisation is characterised by sudden gelation at some point before the reaction is complete. The gel point is observed readily as the time when the mixture suddenly loses fluidity as viscosity rises sharply

$$n \quad \overset{CH_2-CH_2}{\underset{H_2C-CH_2}{\overset{H_2C}{\bigcirc}}}\overset{C=O}{\underset{NH}{}} \quad \longrightarrow \quad [(CH_2)_5CONH]_n \qquad (2.21)$$

and any bubbles present no longer rise through the medium. The gel component at this point represents only a fraction of the total mixture and since it is insoluble in all non-degrading solvents it can be readily freed of the soluble component (the so-called 'sol') by extraction with solvent. As reaction proceeds beyond the gel point so the proportion of gel increases at the expense of the sol.

Gelation occurs at relatively low conversions of monomer to polymer; hence the number-average molar mass at the gel point is low. By contrast, however, the weight-average molar mass becomes infinite at the gel point.

In considering step polymerisation with polyfunctional molecules a number of assumptions are made. They are (i) that all functional groups are equally reactive, (ii) that reactivity is independent of molar mass or solution viscosity, and (iii) that all reactions occur between functional groups on different molecules, *i.e.* there are no intra-molecular reactions. It is found experimentally that these assumptions are not completely valid and tend to lead to an underestimate of the extent of reaction required to bring about gelation.

2.10 Copolymerisation

Step polymerisations tend to be carried out using two different bi-functional molecules so that these give rise to molecules which are essentially copolymers. For example, nylon 6,6 is prepared from hexamethylenediamine and adipic acid; it thus consists of alternating residues along the polymer chain and may be thought of as an al-ternating copolymer.

On the other hand, nylon 6 is prepared from caprolactam, which behaves as a bifunctional monomer bearing two different functional groups, and hence the polymer is made up of just one type of unit along the backbone; it is therefore a homopolymer.

Chain reactions carried out on one type of monomer give rise to homopolymers; when using two types of monomer the situation is more complicated. For example, polymerising mixtures of vinyl chloride with acrylate esters gives rise to a range of molecules, the first of which are relatively rich in acrylate; molecules formed later, when the amount of acrylate monomer is relatively depleted, are richer in vinyl chloride. In a number of instances, reactions of this kind can be used to prepare polymers containing monomers which will not homopolymerise, *e.g.* maleic anhydride and stilbene (vinylbenzene).

In a copolymerisation involving two monomers, A and B, reacting polymer chains can have either monomer A or monomer B at the growing end. As a result, there are four possible reactions: Chain-end A (written $\sim A^\bullet$) with monomer A or monomer B, and chain-end B ($\sim B^\bullet$) with A or with B. Each has a different rate constant, which can be designated k_{AA}, k_{AB}, k_{BA}, and k_{BB} respectively.

Once the copolymerisation reaction has become properly estab-lished, the radical chains $\sim A^\bullet$ and $\sim B^\bullet$ each achieve a steady-state concentration. The absolute rate of conversion of $\sim A^\bullet$ to $\sim B^\bullet$ and of $\sim B^\bullet$ to $\sim A^\bullet$ becomes the same:

$$k_{BA}[\sim B^\bullet][A] + k_{AB}[\sim A^\bullet][B]$$

where brackets represent 'concentration of....'.

The rates of consumption of the two monomers are respectively:

$$-d[A]/dt = k_{AA}[\sim A^\bullet][A] + k_{BA}[\sim B^\bullet][A]$$

and

$$-d[B]/dt = k_{AB}[\sim A^\bullet][B] + k_{BB}[\sim B^\bullet][B]$$

The ratio of the rate constants k_{AA} and k_{AB} can be designated as r_1 and that of k_{BA} to k_{BB} as r_2. Using these so-called *reactivity ratios* and combining the rate equations gives us:

$$\frac{d[A]}{d[B]} = \frac{[A]}{[B]} \cdot \frac{r_1[A] + [B]}{r_2[B] + [A]}$$

This is known as the copolymer equation, and is a mathematical expression from which it is possible to determine the ratio of monomer units in the increment of copolymer formed from a given ratio of monomer molecules.

The value of the reactivity ratios is critical in determining the composition of the copolymer. If the reactivity ratio is greater than 1, the radical prefers to react with chains having the same kind of terminal unit, *e.g.* $\sim A^\bullet$ with A. On the other hand, if the reactivity ratio is less than 1, the monomer prefers to react with chains which end in the other kind of monomer. In the special case that $r_1 r_2 = 1$, the reaction is described as *ideal* copolymerisation because it results in a truly random copolymer whose composition is the same as the composition of the reaction mixture from which polymerisation took place.

Finally where both reactivity ratios take the value of zero, the monomers do not react at all, with growing polymer chains terminated in their own kind of monomer unit. This results in *alternating copolymerisation*. A few typical monomer reactivity ratios are given in Table 2.3.

In addition to the types of copolymer already mentioned there are two other important classes of copolymer, namely block and graft copolymers.

Block copolymers are those containing long sequences of the same monomer unit along the backbone:

—AAAA—BBBB—AAAA—

Methods of synthesising these polymers are available and may include polymerisation of low molar mass units ('blocks') of homopolymer which are later reacted together to yield the copolymer.

Table 2.3 Typical monomer reactivity ratios[a] (reaction temperature 60 °C in each case).

Monomer 1	Monomer 2	r_1	r_2
Acrylonitrile	Styrene	0.04	0.40
Methyl methacrylate	Styrene	0.46	0.52
Styrene	Vinyl chloride	17	0.02
Styrene	Vinyl acetate	55	0.01

[a]Data from L. J. Young, Copolymerisation Reactivity Ratios, in *Polymer Handbook*, ed. J. Brandrup and E. H. Immergut, Wiley-Interscience, New York, 2nd edn, 1975, pp. II-105–386.

Graft copolymers are those in which a homopolymer backbone has a number of branches, which are themselves homopolymers of another monomer, grafted on:

—AAAAAAAAA—
B
B
B
B

3 Polymer Structure

This chapter is concerned with aspects of the structure of polymeric materials outside those of simple chemical composition. The main topics covered are polymer stereochemistry, crystallinity, and the character of amorphous polymers including the glass transition. These may be thought of as arising from the primary structure of the constituent molecules in ways that will become clearer as the chapter progresses.

Before proceeding, a word on nomenclature is necessary. Polymer chemists, following the example of P. J. Flory, have tended to use the words configuration and conformation in a sense that differs from that conventionally employed within organic chemistry. In this book, by contrast, I intend to go along with F. W. Billmeyer, and use these words in the way that they apply more widely throughout chemistry. Thus *configuration* is the term given to an arrangement of atoms that cannot be altered except by breaking chemical bonds, while *conformation* is the term applied to the individual, recognisable arrangement of atoms that can be altered by simple rotation around a single bond. Configurations include head-to-tail arrangements, described in the previous chapter; conformations include *trans versus gauche* arrangements of successive carbon-carbon bonds along the backbone of an individual macromolecule.

3.1 Polymer Stereochemistry

Monosubstituted vinyl monomers form polymers containing a series of asymmetric carbon atoms along the molecule. The precise

The Chemistry of Polymers, 5th Edition
By John W. Nicholson
© John W. Nicholson 2017
Published by the Royal Society of Chemistry, www.rsc.org

arrangement of these asymmetric carbons gives rise to three different possible stereochemical arrangements. Firstly, where all asymmetric carbon atoms adopt identical configurations, the resulting polymer is described as *isotactic*. Secondly, where there is a regular alternating arrangement of asymmetric carbon atoms, the polymer is described as *syndiotactic*. Lastly, where there is no regularity at all in the arrangement of asymmetric carbon atoms, the resulting structure is known as *atactic*. These three steric arrangements are illustrated in Figure 3.1.

The synthesis of isotactic and syndiotactic polymers has been achieved for a number of polymers. For example poly(methyl methacrylate) can be prepared in either isotactic or syndiotactic configurations depending on the details of the polymerisation conditions.

Polymer stereochemistry, sometimes referred to as tacticity, is not the only source of variation in polymer configuration. For the mono-substituted butadiene isoprene, the structures shown in Figure 3.2 are possible.

Polymers containing each of these configurations are known, the most common being the *cis*-1,4- and the *trans*-1,4-isomers. The first of these, poly(*cis*-1,4-isoprene), is the macromolecular constituent of natural rubber; the second is the material known as gutta percha. The latter, unlike natural rubber, has no elastomeric properties, but has a leathery texture. It has been used for diverse applications such as

Figure 3.1 Possible stereochemical arrangements of polymers containing asymmetric carbon atoms.

Figure 3.2 Isomers of poly(isoprene).

golf-ball covers and as an insulating material for the *trans*-Atlantic cables of the late nineteenth century.

3.2 Polymer Crystallinity

The essential requirement for crystallinity in polymers is some sort of stereoregularity. This is not to say that the entire collection of macromolecules within the sample needs to be all isotactic or syndiotactic; however, it is essential that regions along the backbone of a significant number of the macromolecules do have such regularity.

Many polymers show partial crystallinity. This is apparent from the study of X-ray diffraction patterns, which for polymers generally show both the sharp features associated with crystalline regions as well as less well-defined features which are characteristic of disordered substances with liquid-like arrangements of molecules. The co-existence of crystalline and amorphous regions is typical of the behaviour of 'crystalline' polymers.

The reason for the existence of both phases in polymers is quite straightforward. Crystalline regions are formed from the stereoregular components of the macromolecules, which, given the nature of the polymerisation processes from which the macromolecules are built up, represent only a proportion of the overall material. There is often the possibility that an essentially isotactic polymer will contain a syndiotactic segment and *vice versa*. Chain branching may also occur, thereby reducing local regularity in structure and inhibiting crystallisation. Occasionally copolymerisation is the reason for the co-existence of the amorphous regions with the crystalline phases. For example, the inclusion of a small amount of isoprene in what is mainly poly(isobutylene), so-called butyl rubber, limits the extent of possible crystallisation but does not prevent its occurrence completely.

Non-crystalline polymers are those which include high levels of irregularity within their structure. Typical sources of such irregularity are copolymerisation with significant amounts of at least two co-monomers and also complete absence of stereoregularity, *i.e.* atactic polymers.

Among the polymers which contain clear crystalline phases are poly(ethylene) and PTFE. Their properties, however, are very different from those usually associated with crystalline solids; in particular, they tend to exhibit low rigidity and at ambient temperatures are soft and deformable rather than brittle. This is because, unlike say sodium chloride or copper sulphate, crystalline polymers include both amorphous and crystalline regions.

X-Ray studies of crystalline polymers have shown that the crystallites are very small, of the order of tens of nanometres. This is much less than the end-to-end distance of a high molar mass polymer, a fact which was originally interpreted as implying that an individual polymer molecule actually passed through several crystallites as well as through a number of amorphous regions. The process of crystallisation was viewed as occurring by the coming together of bundles of regular segments from different molecules to form a close packed crystalline array at localised points within the material (see Figure 3.3a). This theory, which is to be found described in older books and papers in polymer science, is known as the fringed micelle theory.

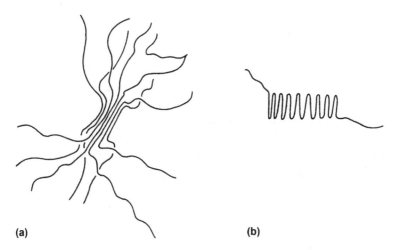

(a) (b)

Figure 3.3 Arrangements of molecules in crystalline polymers according to (a) fringed micelle and (b) lamellae theories.

The fringed micelle theory has been less favoured recently following research on the subject of polymer single crystals. This work has led to the suggestion that polymer crystallisation takes place by single molecules folding themselves at intervals of about 10 nm to form lamellae as shown in Figure 3.3b. These lamellae appear to be the fundamental structures of crystalline polymers.

It has proved difficult to decide which of these two theories of polymer crystallisation is correct, since both are consistent with the observed effects of crystallinity in polymers. These effects include increased density, increased stiffness, and higher softening point. However, the balance of opinion among those working with crystalline polymers favours the latter theory, based on lamellae formed by the folding of single molecules.

3.2.1 Orientation and Crystallisation

When a rubbery polymer, such as natural rubber, is stretched the molecules become aligned. This orientation leads to crystallisation. The effect of this so-called strain-induced crystallisation is to make the extended polymer stiffer than the unstrained polymer. Such crystallisation is not permanent but disappears when the sample is allowed to retract and regain its original dimensions.

Other, non-rubbery polymers can be made to increase their crystalline proportion permanently by stretching under appropriate conditions. For example, heating nylon 6,6 to a temperature between its transition temperature and the melting point of the crystalline regions followed by stretching causes additional crystallisation with the crystallites being aligned with the direction of the extension. The oriented crystalline filaments or fibres that result are much stronger than the unoriented material and may approach 700 MPa in strength.

Similarly, oriented crystallisation can be induced by stretching sheets or films of polymers in two directions simultaneously. The resulting materials have biaxially oriented polymer crystals. Typical examples of such materials are biaxially stretched poly(ethylene terephthalate), poly(vinylidene chloride), and poly(propylene). Since the oriented crystals do not interfere with light waves, such films combine good strength with high clarity, which makes them attractive in a number of applications.

3.2.2 The Crystalline Melting Point

Unlike conventional organic substances, crystalline polymers do not have well-defined melting points. This is because they are effectively mixtures, comprising components of a range of relative molar masses, the individual components of which melt at different temperatures. Thus low relative molar mass homopolymers melt at lower temperatures than high relative molar mass species. The size of the crystallites themselves also influences the range of the melting temperature, since smaller and less perfect crystallites melt before larger ones as the temperature is raised.

Other factors which influence the value of the range of temperature over which melting takes place are (*a*) the presence of co-monomers in the polymer and (*b*) the presence of traces of solvent or plasticiser in the polymer. The first of these, involving copolymerisation, lowers the melting point by shortening the length of crystallisable sequences within the individual polymer molecules. The latter does so by increasing the relative mobility of the polymer molecules in the material, thereby reducing the energy necessary to take them into the liquid phase.

3.2.3 Polyblends

In the design of new materials there is currently less emphasis on producing new polymers than on producing novel combinations of polymers by blending, so-called polyblends. The polymers involved in polyblends generally have very different properties so that typically a brittle, glassy polymer would be blended with a rubbery polymer, thereby producing a material of good rigidity *and* toughness. Such a combination of properties tends to be lacking in most polymers. A good example of a polyblend is high-impact poly(styrene), in which rubber particles are blended with the base polymer, as described in Chapter 1.

The morphology of polyblends tends to be the same as that shown by high-impact poly(styrene), *i.e.* a dispersion of rubbery particles in a matrix of the glassy component. This kind of arrangement is responsible for the improvement in mechanical properties observed. One particular change associated with the polyblends is that a significant difference develops between the crazing stress and the fracture stress. In glassy polymers with no rubbery filler, crazing occurs just before fracture, and the stresses are very close in value. In polyblends, crazing occurs well before the sample actually undergoes fracture. This phenomenon appears to occur because the rubber particles are large enough to bridge the craze in the glassy matrix and become load bearing, as illustrated in Figure 3.4. Such a mechanism explains the fact that there is an optimum size range for the rubber particles which bring about the increase in fracture strength. For

Figure 3.4 Mechanism of reinforcement by rubber particles in a glassy polymer matrix.

example, for high impact poly(styrene) the best results are obtained with rubber particles in the size range 1–10 μm.

3.3 Thermal and Mechanical Properties

Three factors affect the essential nature of a polymeric material and determine whether it is glassy, rubbery, or fibre-forming under a given set of conditions. These are:

   (i) the flexibility of the macromolecule,
  (ii) the magnitude of the forces between the molecules, and
 (iii) the stereoregularity of the macromolecules.

These three factors influence the ability of the polymer to crystallise, the melting point of any resulting crystalline regions, and also the glass transition temperature. It is the last of these features of polymeric materials which we will concentrate on for the rest of this chapter.

3.3.1 The Glass Transition Temperature, T_g

The glass transition is a phenomenon observed in linear amorphous polymers, such as poly(styrene) or poly(methyl methacrylate). It occurs at a fairly well-defined temperature when the bulk material ceases to be brittle and glassy in character and becomes less rigid and more rubbery.

Many physical properties change profoundly at the glass transition temperature, including coefficient of thermal expansion, heat capacity, refractive index, mechanical damping, and electrical properties. All of these are dependent on the relative degree of freedom for molecular motion within a given polymeric material and each can be used to monitor the point at which the glass transition occurs.

Unfortunately, in certain cases, the values obtained from these various techniques can vary widely. An example is the variation found in the measured values of T_g for poly(methyl methacrylate), which range from 110 °C using dilatometry (*i.e.* where volume changes are monitored) to 160 °C using a rebound elasticity technique. This, though, is an extreme example; despite the fact that the measured value of T_g does vary according to the technique used to evaluate it, the variation tends to be over a fairly small temperature range.

The glass transition is a second-order transition. In this it differs from genuine phase changes that substances may undergo, such as melting or boiling, which are known as first-order transitions. These latter transitions are characterised by a distinct volume change, by changes in optical properties (*i.e.* in the X-ray diffraction pattern and the infrared spectrum) and by the existence of a latent heat for the phase change in question. By contrast, no such changes occur at the glass transition, though the rate of change of volume with temperature alters at the T_g, as illustrated in Figure 3.5.

The glass transition can be understood by considering the nature of the changes that occur at the temperature in question. As a material is heated to this point and beyond, molecular rotation around single bonds suddenly becomes significantly easier. A number of factors can affect the ease with which such molecular rotation takes place, and hence influence the actual value that the glass transition temperature takes. The inherent mobility of a single polymer molecule is important and molecular features which either increase or reduce this mobility will cause differences in the value of T_g. In addition, interactions between polymer molecules can lead to restrictions in molecular mobility, thus altering the T_g of the resulting material.

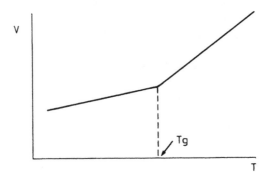

Figure 3.5 Plot of volume against temperature for a typical polymer passing through its glass transition.

Briefly, the following features are known to influence the glass transition temperature:

(a) The presence of groups pendant to the polymer backbone, since they increase the energy required to rotate the molecule about primary bonds in the main polymer chain. This is especially true of side chains or branches.
(b) The presence of inherently rigid structures in the backbone of the molecule, *e.g.* phenylene groups.
(c) Crosslinking.
(d) Hydrogen bonds between polymer chains.
(e) Relative molar mass, which influences T_g because higher molar mass polymers have less ease of movement and more restrictions in their overall molecular freedom than polymers of lower molar mass.
(f) The presence of plasticisers. These are discussed in detail in the next section of this chapter.

The effects of these different factors can be seen in the T_g values of some typical polymers. A number of these values are shown in Table 3.1, together with a brief note about what feature particularly contributes to the relative level of the glass transition temperature.

The glass transition temperature is usually insensitive to the relative molar mass of the polymer above degrees of polymerisation of about 200. Below this, T_g is related to relative molar mass, lower molar mass specimens having lower values of T_g. This is because the chain ends tend to have inherently greater degrees of freedom of motion than the other segments of the macromolecule, and hence the more chain ends per unit volume of material, the greater the overall freedom of the molecules, leading to reduced values of T_g.

Random copolymers having an amorphous morphology generally exhibit a single T_g having a value between the T_gs of the individual homopolymers. The exact value depends on the relative proportions of the respective monomers in the copolymer.

Table 3.1 Glass transition temperatures of some typical polymers.

Polymer	T_g (°C)	Contributing feature
Poly(ethylene)	-20	Flexible backbone
Poly(propylene)	$+5$	CH_3 group inhibits freedom of rotation
PVC	$+80$	Strong polar attraction between molecules
PTFE	$+115$	Very stiff backbone

In certain cases polyblends of two very compatible polymers also show just one T_g. The value of such a T_g can be predicted from either of the following equations:

$$(T_g)_{AB} = (T_g)_A f_A + (T_g)_B f_B \qquad (3.1)$$

where f_A and f_B are the volume fractions of A and B respectively, or

$$\frac{1}{T_g} = \frac{M_A}{(T_g)_A} + \frac{M_B}{(T_g)_B} \qquad (3.2)$$

where M_A and M_B are the mass fractions of A and B.

The latter equation may also be used to predict the approximate T_g of a random copolymer, using the mass fractions of the respective monomers from which the copolymer was prepared.

Both of these are equations are approximate and are useful only for giving estimates of the value of the T_g of the polyblend or copolymer. To calculate values of T_g more accurately requires additional information such as the coefficients of thermal expansion of both components in both their liquid and glassy states. Given the uncertainty in the numerical value of T_g, which as we have seen depends on the method by which T_g has been determined, there is little point in developing such arithmetical refinements.

More usually, as has been described earlier, polyblends are made from incompatible polymers that give a two-phase structure. These polyblends show two T_gs, one for each phase. The temperatures of these transitions correspond closely to the T_gs of the respective homopolymers.

3.3.2 The Effect of Plasticisers

Plasticisers are substances with relative molar masses well below those of polymers, usually liquids that, when added to polymers, give apparently homogeneous materials that are softer, more flexible, and easier to process than the polymer alone. A typical example of the use of plasticisers is the incorporation of di-isooctyl phthalate into PVC, which at levels of 70 parts of phthalate to 100 parts of PVC converts the polymer from a rigid solid at room temperature into a rubbery material with uses in the field of protective clothing. Thus the T_g of the PVC is reduced from a value higher than room temperature to one that occurs below it.

Compared with conventional organic solvents, plasticisers have generally high relative molar masses, typically in excess of 300. They are essentially non-volatile solvents which interact with the polymer molecules in the same way as conventional solvents. However, they are used at such levels that there is an excess of polymer. Hence the plasticiser is absorbed by the polymer to form a material having a lower T_g than the polymer alone. Because of the high relative molar mass of the substances used as plasticisers, they diffuse into the polymer only very slowly at room temperature. To overcome this difficulty they are generally blended with the polymer at elevated temperatures. Alternatively, they may be incorporated by mixing them with the polymer in the presence of lower relative molar mass solvents which are volatile and can be removed at a later stage in the processing.

The main mode of action of plasticisers appears to be to act as spacers at the molecular level. Hence less energy is required to free the molecules sufficiently to allow substantial rotation about the C–C bonds; thus T_g is lowered. So, too, is the temperature at which the polymer begins to flow.

3.3.3 Methods of Determining Glass Transition Temperature

A number of methods exist for determining T_g. For example, the change in volume as the temperature is raised can be measured dilatometrically and the slope of plot of volume against temperature recorded. As we have seen from Figure 3.5, such a plot shows a distinct change at T_g. Alternatively, Young's modulus, the ratio of stress to strain, can be measured at increasing temperatures. Around the glass transition there is a dramatic drop in Young's modulus corresponding to several orders of magnitude (see Figure 3.6).

Differential Scanning Calorimetry (DSC) is a widely used technique for determining T_g. In DSC, heat is supplied to an experimental sample and a reference material with the aim of raising the temperature of each at uniform rates. This is shown on a plot as a straight, horizontal line. When a transition occurs in the sample, the rate of heating required by the two materials becomes different, and this shows in a change from the horizontal line. For a transition such as melting, the change appears as a peak in the DSC trace. For the T_g, which is a secondary transition, there is not a peak but an upward slope in the trace. Once past the transition, the trace continues as a horizontal line, though off-set compared with the earlier part. The glass transition is thus recorded. However, determining the precise point at which it begins is difficult, so it is usually determined from

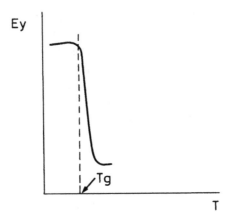

Figure 3.6 Plot of Young's modulus against temperature for a typical polymer passing through its glass transition.

the "half-height" method, whereby the half-way point on the upward slope is determined, and taken to be the T_g.

Finally, there is the extremely important group of *relaxation* methods for determining T_g. These can be based on either mechanical (sometimes thermo-mechanical) or electrical relaxations occurring within the material, and, although they do not always give results that are completely consistent with those obtained by the static mechanical tests already mentioned, they are considered very reliable and are widely used.

The relaxation methods employed are Dynamic Mechanical Thermal Analysis (DMTA) and Dielectric Thermal Analysis (DETA). Generally in both cases a single excitation frequency is used and the temperature is varied, typically over a range between -100 °C and $+200$ °C. Changes in molecular motion, and hence T_g, are detected by both techniques, but in the case of DETA the process has to involve movement of dipoles or fully developed electrical charges on the polymer in order to be detected. Thus the two techniques can be used to complement each other, since transitions can be detected on DMTA and assigned as due to dipoles according to whether or not they also occur with DETA.

Dynamic mechanical techniques for studying polymers are described in detail in Chapter 7. For the moment we will restrict ourselves to a simple outline of the method of DMTA as it is applied to the determination of T_g.

Experimentally DMTA is carried out on a small specimen of polymer held in a temperature-controlled chamber. The specimen is subjected to a sinusoidal mechanical loading (stress), which induces a corresponding extension (strain) in the material. The technique of DMTA essentially uses these measurements to evaluate a property known as the complex dynamic modulus, E^*, which is resolved into two component parts, the storage modulus, E', and the loss modulus, E''. Mathematically these moduli are out of phase by an angle δ, the ratio of these moduli being defined as tan δ, *i.e.*

$$\tan \delta = \frac{E''}{E'} \tag{3.3}$$

The numerical value of tan δ varies with temperature and reaches a maximum at T_g, after which it rapidly falls, making the position of the maximum very distinct. Hence determining T_g from a typical DMTA spectrum is fairly straightforward, as shown in the example in Figure 3.7.

Details of the technique can be varied, and oscillation modes involving torsion, single and dual cantilever, tension, compression and 3-point bend can all be employed. Heating rates, oscillation frequencies and strains can also all be varied. DMTA is very sensitive and

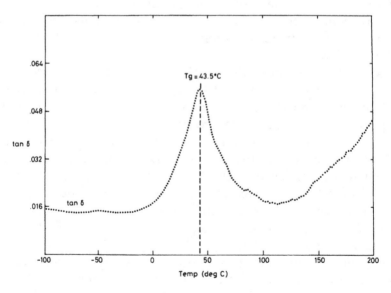

Figure 3.7 DMTA spectrum for an alkyd paint film attached to an aluminium substrate, $T_g = 43.5$ °C.
(From G. J. Bratton, E. A. Wasson, J. W. Nicholson and A. D. Wilson, *J. Oil Colour Chem. Assoc.*, **72**, 1989, 10).

can be used for polymers, such as epoxies, where determination of T_g is difficult.

By contrast, the technique of DETA is based on electrical measurements. In this technique polymers are exposed to an alternating current and changes in conductance (G) and capacitance (C) with increasing temperature are measured. By definition:

$$\tan \delta_D = \frac{G}{2\pi Cf} \tag{3.4}$$

where f is the applied frequency.

As in DMTA, values of $\tan \delta_D$ determined by DETA reach a maximum at T_g. Hence for polymers containing groups sensitive to external electrical fields DETA is a convenient method of determining T_g. In addition, as previously mentioned, using DETA in conjunction with DMTA can be useful for determining whether the glass transition involves polar or charged groups in the polymer.

3.3.4 The Effect of Polymer Stereochemistry on T_g

As a general principle the value taken by T_g varies according to the stereochemistry of the macromolecules in the order:

synodiotactic > atactic > isotactic

Such a sequence is very pronounced with the different stereoisomers of poly(methyl methacrylate), for example, and is a reflection of the increasingly easy rotation about the backbone C–C bonds in the different configurations.

Similarly, there are differences between the values of T_g for *cis-* and *trans-*isomers in a polymer such as 1,4-poly(isoprene). This is an area of some scientific debate, because early data, which are now considered unreliable, showed that the T_g of the *transisomer* was -53 °C compared with -73 °C for the *cis-*form. More recent studies using the technique of Differential Scanning Calorimetry have shown instead that the values of T_g for these isomers are much closer together, with the *trans-form* having a slightly lower value of T_g than the *cis* (-70 °C compared with -67 °C). In part, though, these problems in understanding the effects of polymer stereochemistry arise because of experimental difficulties in determining T_g, and, as we have seen earlier, from the fact that the actual value obtained for T_g depends on which technique has been used to determine it.

3.3.5 The Relationship between Crystalline Melting Point and T_g

As was mentioned previously, crystalline polymers always contain a significant amorphous region, which is capable of undergoing a glass transition. Such transitions are undergone only by amorphous phases and always at temperatures below the crystalline melting point, T_m. Thus heating a crystalline polymer causes it to pass first through the second-order transition at T_g, and then through a first-order transition as the crystallites undergo a true phase change on melting at T_m.

Features of chemical structure that affect the degree of molecular freedom influence both the crystalline melting point and the glass transition temperature. Moreover, such features have roughly similar effects on both properties, so that the empirical rule has been found that for many polymers:

$$T_g = 0.66 \pm 0.44 \ T_m \qquad (3.5)$$

where temperatures are given in Kelvin.

This rule of thumb does not apply to all polymers. For certain polymers, such as poly(propylene), the relationship is complicated because the value of T_g itself is raised when some of the crystalline phase is present. This is because the morphology of poly(propylene) is such that the amorphous regions are relatively small and frequently interrupted by crystallites. In such a structure there are significant constraints on the freedom of rotation in an individual molecule which becomes effectively tied down in places by the crystallites. The reduction in total chain mobility as crystallisation develops has the effect of raising the T_g of the amorphous regions. By contrast, in polymers that do not show this shift in T_g the degree of freedom in the amorphous sections remains unaffected by the presence of crystallites, because they are more widely spaced. In these polymers the crystallites behave more like inert fillers in an otherwise unaffected matrix.

This part of the chapter has shown that the relationship between T_g and T_m is complicated. This relationship needs to considered separately for each polymer, but can be useful for gaining an insight into the morphology of particular semi-crystalline polymers.

3.3.6 Other Thermal Transitions

This chapter has concentrated on one of the most important of the transitions undergone by polymeric materials as their temperature is raised, namely the glass transition. However, this is not the only transition of this type undergone by polymers. In some materials there are a number of other second-order transitions of a similar nature to the glass transition. They occur because side chains and small segments of the backbone of the molecule require less energy for mobility than the main segments associated with the glass transition. Typically these transitions are labelled in descending order of temperature α, β, γ, and so on. The T_g normally corresponds to the transition. These labels do not imply any similarity in origin of the transition so that, for example, the α-transition in one polymer may arise from a different molecular motion from the α-transition in a second polymer.

4 Crosslinking

4.1 Introduction

Covalent chemical bonds that occur *between* macromolecules are known as *crosslinks*. Their presence and density have a profound influence on both the chemical and mechanical properties of the materials in which they occur.

As we have seen previously the presence of crosslinks between macro-molecules influences the way in which these materials respond to heat. Uncrosslinked polymers will generally melt and flow at sufficiently high temperatures; they are usually thermoplastic. By contrast, crosslinked polymers cannot melt because of the constraints on molecular motion introduced by the crosslinks. Instead, at temperatures well above those at which thermoplastics typically melt, they begin to undergo irreversible degradation.

Response to heat is not the only difference in chemical behaviour between crosslinked and uncrosslinked polymers. Dissolution behaviour is also different since this, too, depends on the nature and extent of any interchain covalent bonds. An uncrosslinked polymer will usually dissolve in an appropriate solvent. The process may well be lengthy, but given sufficient time and adequate polymer–solvent compatibility, the polymer will dissolve. By contrast, crosslinked polymers will not dissolve. Solvation of chain segments cannot overcome the effect of the covalent bonds between the macromolecules, hence crosslinked molecules cannot be carried off into the solution. Depending on the crosslink density, however, such materials may admit significant amounts of solvent, becoming softer and swollen as

The Chemistry of Polymers, 5th Edition
By John W. Nicholson
© John W. Nicholson 2017
Published by the Royal Society of Chemistry, www.rsc.org

they do so. Such swelling by fairly lightly crosslinked materials is generally reversible and, given appropriate conditions, solvent that has entered a crosslinked structure can be removed and the polymer returned to its original size.

The mechanical properties of polymers also depend on the extent of crosslinking. Uncrosslinked or lightly crosslinked materials tend to be soft and reasonably flexible, particularly above the glass transition temperature. Heavily crosslinked polymers, by contrast, tend to be very brittle and, unlike thermoplastics, this brittleness cannot be altered much by heating. Heavily crosslinked materials have a dense three-dimensional network of covalent bonds in them, with little freedom for motion by the individual segments of the molecules involved in such structures. Hence there is no mechanism available to allow the material to take up the stress, with the result that it fails catastrophically at a given load with minimal deformation.

Again because of the crosslinks, such brittle behaviour occurs whatever the temperature; unlike brittle materials based on linear polymers, there is no temperature at which molecular motion is suddenly freed. In other words, the T_g, if there is one, does not produce dramatic changes in mechanical properties so that the material is changed from one that undergoes brittle behaviour to one that exhibits so-called tough behaviour.

Crosslinking can be introduced into an assembly of polymer molecules either as the polymerisation takes place or as a separate step after the initial macro-molecule has been formed. Typical of the first category are polymers made by step processes, often condensations, in which monomers of functionality greater than 2 are present. The relative concentration of such higher functionality monomers then determines the density of crosslinks in the final material.

The second category involves the formation of a prepolymer followed by crosslinking. Because the second step can be considered as correcting faults in the material, such as a tendency to flow at unacceptably low temperatures, this second step is often referred to as 'curing'. There are examples of such curing involving both chain and step reactions. For instance, step processes can be carried out with an excess of one kind of higher functionality monomers, perhaps hydroxy-groups. The initial polymer may be essentially linear, but contains numerous hydroxy-groups through which crosslinking can be made to occur at a later stage when sufficient complementary monomer, perhaps a bifunctional acid, has been added to the mixture.

Certain commercially important crosslinking reactions are carried out with unsaturated polymers. For example, as will be described later in this chapter, polyesters can be made using bifunctional acids which contain a double bond. The resulting polymers have such double bonds at regular intervals along the backbone. These sites of unsaturation are then crosslinked by reaction with styrene monomer in a free-radical chain (addition) process to give a material consisting of polymer backbones and poly(styrene) co-polymer crosslinks.

Another commercially important crosslinking process that involves unsaturated polymer precursors is the so-called 'drying' of alkyd resins in paints. This process is not drying at all, at least not in the sense of mere loss of solvent to leave behind a solid residue. Instead, the main process is the conversion of high relative molar mass molecules to a crosslinked structure *via* reaction with, and crosslinking by, atmospheric oxygen. The process is complex, but the outlines are known. Briefly, aided by metal drier catalysts, oxygen seems to react with a methylene group adjacent to a double bond in the oil molecule to form a hydroperoxide (Reaction (4.1)). This is followed by a shift in the position of the double bond to form conjugated structures; often, there is a change in configuration from *cis* to *trans*. Next, the hydroperoxide groups break down to yield free radicals, the products of which are mixtures of types of crosslink, including R–R, R–O–R, and R–O–O–R. In generating such species the relative molar mass progressively increases and the formerly liquid oil solidifies to a soft, elastic material able to retain adhesion to substrates as they expand and contract during natural fluctuations of temperature.

$$RH \quad + \quad O_2 \quad \longrightarrow \quad ROOH \qquad (4.1)$$

In the sections that follow a number of other important crosslinking processes are described. Since each of these processes is of commercial importance control of the final product is essential. Hence these crosslinking reactions tend to be carried out in such a way that they are essentially two-stage, with a pre-polymer formed initially, followed by the crosslinking step. This is despite the fact that, as will be seen, in at least one of the four cases the chemistry of the polymerisation is such that there is an in-built tendency to give three-dimensional networks in the early stages of reaction. However, by controlling the ratio of reactants carefully, early crosslinking is avoided and greater control is exerted over the final step.

4.2 Phenol–Formaldehyde Resins

Crosslinking in phenol–formaldehyde resins is carried out on essentially linear prepolymers which have been formed by having one of the components in sufficient excess to minimise crosslinking during the initial step. These prepolymers may be one of two kinds: the so-called resoles or the so-called novolaks.

Resole resins are formed by having formaldehyde in excess, usually in mole ratio of about 2 : 1 on the phenol. By contrast, novolak resins are formed by having phenol in excess, generally in mole ratio of 1.25 : 1 on formaldehyde. The pH at which these prepolymers are fabricated is also different; resoles are formed under alkaline conditions and the novolaks under acidic conditions.

The route to crosslinked phenol–formaldehyde resins *via* resoles corresponds to that used by Baekeland in his original commercial technique. They now tend to be used for adhesives, binders, and laminates. The resole prepolymers are made typically in batch processes, using a trace of ammonia (about 2% on phenol) as the alkaline catalyst. Care has to be taken with this process since, despite the molar excess of formaldehyde, there is sufficient of each component present in the prepolymer to permit the formation of a highly crosslinked product. Indeed, such a product will form if the resole is heated excessively, but the problem can be avoided by careful attention to the conditions of reaction and by ensuring that polymerisation is not allowed to proceed for too long.

Novolak resins cannot be converted into a network structure on their own but require the addition of some sort of crosslinking agent. In principle additional formaldehyde can be used, but in practice this is not the substance chosen. Instead, hexamethylenetetramine (HMT) is used. This substance is prepared by the interaction of formaldehyde with ammonia as illustrated in Reaction (4.2). This increases the complexity of the reactions that occur on crosslinking with the result that the curing of novolaks with HMT is far from properly understood. However, the final material is known to consist of phenol rings joined mainly by methylene groups with a small number of different nitrogen-containing links.

The crosslinking of resoles is slightly more straightforward than that of novolaks, if only because there are fewer possible chemical structures involved in the setting reaction or finished structure. Since resoles are prepared under alkaline conditions, crosslinking is generally preceded by neutralisation. This enhances the ease with which network structures can form when the resin is simply heated.

$$4\,NH_3 \quad + \quad 6\,CH_2O \longrightarrow \qquad\qquad\qquad \tag{4.2}$$

The setting chemistry of resole resins is complex, and experimentally difficult to study, mainly because the cured product, being insoluble, is not amenable to ready chemical investigation. Part of the information on these materials has come from studies of model systems such as mononuclear methylphenols, which give soluble products. These products present fewer difficulties in chemical analysis.

An early process in the cure reaction is protonation of a methylolphenol, followed by loss of a molecule of water to produce a benzylic carbonium ion (see Reaction (4.3)). This may be followed by reaction with a second phenol to generate a bridged structure, as illustrated in Reaction (4.4). Alternatively the benzylic carbonium atom may react with another methylol group, thus generating a bridge based on an ether-type structure (see Reaction (4.5)).

$$\tag{4.3}$$

$$\tag{4.4}$$

(4.5)

By these two processes a network is built up that consists of both methylene and ether bridges; in commercial materials there is some evidence that methylene groups predominate.

In addition to the crosslinks already mentioned, there exist a number of possibilities for secondary reactions, so that the final resin may consist of a number of crosslinks whose origin is not immediately clear. For example, this is evidence that some resins contain units that are effectively derivatives of stilbene (vinylbenzene), implying that there are units in the cured resin corresponding to that illustrated (4.1).

(4.1)

From this brief discussion it is clear that crosslinking in phenol–formaldehyde resins is complicated and no individual specimen of these materials can be characterised well at the molecular level. Crosslinking is irregular and variable, though it gives rise to a material having sufficiently acceptable properties that it became the first commercially important plastic material; indeed, as mentioned in Chapter 1, these resins continue to retain some commercial importance in certain specialised applications.

4.3 Unsaturated Polyester Resins

Polyesters are widely used as laminating resins. The prepolymers for these crosslinked materials are viscous pale yellow coloured liquids of

Figure 4.1 Structures of (a) maleic acid and (b) phthalic acid.

fairly low relative molar mass, typically about 2000. They are prepared in a step polymerisation process from a glycol, typically 1,2-propylene glycol (1,2-propanediol), together with both a saturated and an unsaturated dicarboxylic acid. The unsaturated acid provides sites for crosslinking along the backbone, while the saturated acids are present effectively to limit the number of crosslinking points. The effect of limiting these sites is to reduce the brittleness of the cured resin.

There are numerous possible monomers that can be used in the backbone of the polyester prepolymer. However, typical monomers are 1,2-propanediol, as just mentioned, with maleic acid (usually added as the anhydride) to provide the sites of unsaturation, and phthalic acid (again usually added as the anhydride) to act as the second of the two diacid species. The structures of these latter two substances are shown in Figure 4.1.

A number of other monomers may be employed as variations on the materials mentioned so far, to introduce specific properties into the finished resin. For example, halogenated molecules containing either chlorine or bromine atoms may be used to confer fire resistance. As described in Chapter 8, the effect of halogens in the polymer structure is to make the resins difficult to ignite and unable to sustain combustion.

For the preparation of the prepolymer, a mixture corresponding typically to mole ratios of diol: unsaturated acid: saturated acid of 3.3 : 2 : 1 is polymerised by stirring the ingredients together at elevated temperatures, normally 150–200 °C. The slight excess of diol is to allow for possible evaporation losses. The polymerisation reaction requires several hours to proceed sufficiently to give a prepolymer of acceptable relative molar mass.

Having made the prepolymer, a reactive diluent, usually styrene, is added, together with an appropriate free-radical initiator system in order to bring about crosslinking. This cure reaction can be made to occur at either ambient or elevated temperatures and, depending on conditions, is complete in anything from a few minutes to several hours.

In order to bring about crosslinking of polyesters with styrene one of two types of initiator systems is used, which differ in the temperature at which they are effective. For curing at elevated temperatures, peroxides are used which decompose thermally to yield free radicals. Among those peroxides employed are benzoyl peroxide, 2,4-dichlorobenzoyl peroxide, di-t-butyl peroxide, and dodecyl peroxide. Mixtures of polyester prepolymer, styrene, and such initiators are reasonably stable at room temperatures but undergo fairly rapid crosslinking at temperatures between 70 °C and 150 °C, depending on which particular peroxide is used.

For a number of applications curing at room temperature is desirable. This so-called 'cold cure' is brought about by using a peroxy initiator in conjunction with some kind of activator substance. The peroxy compounds in these cases are substances such as methyl ethyl ketone peroxide and cyclohexanone peroxide, which as used in commercial systems tend not to be particularly pure, but instead are usually mixtures of peroxides and hydroperoxides corresponding in composition approximately to that of the respective nominal compounds. Activators are generally salts of metals capable of undergoing oxidation/reduction reactions very readily. A typical salt for this purpose is cobalt naphthenate, which undergoes the kind of reactions illustrated in Reactions (4.6) and (4.7).

$$ROOH \ + \ Co^{2+} \ \longrightarrow \ RO\cdot \ + \ OH^- \ + \ Co^{3+} \tag{4.6}$$

$$ROOH \ + \ Co^{3+} \ \longrightarrow \ ROO\cdot + \ OH^- \ + \ Co^{2+} \tag{4.7}$$

Cold-cure crosslinked polyester resins are used extensively in the production of large glass-fibre reinforced items, which are usually made by the hand lay-up technique. The glass fibres are usually held together in the form of a chopped strand mat. Such a mat consists of chopped lengths of glass fibre about 5 cm long held together by a resinous binder. The glass fibers may be treated with some kind of silicone finish containing unsaturated organic groups. This aids wetting of the fibres on mixing with uncured resin and strengthens the bond between the matrix and the fibres, thereby improving the mechanical properties of the cured material.

When cured, glass-fibre reinforced resins are strong, yet lightweight, with typical properties such as those illustrated in Table 4.1. These materials are prone to attack by alkaline environments because of the susceptibility of the ester groups within them to hydrolysis. Away from alkaline environments they have excellent weathering characteristics and are widely used to construct the hulls of boats, roofing panels, sports car bodies, swimming pool liners, and tanks and storage vessels in chemical plants.

Table 4.1 Typical properties of hand lay-up glass-mat reinforced polyester resin.

Property	Typical values
Specific gravity	1.4–1.5
Tensile strength/MPa	8–17
Flexural strength/MPa	55–117
Water absorption/%	0.2–0.8

4.4   Polyurethanes

Although the name polyurethane might be taken as implying that these materials contain urethane groups (–NHCOO–) in the backbone of the macro-molecule, for those polyurethanes in major commercial use this is not true. For such materials the initial macromolecule tends to be a polyester or polyether; it is the crosslinks that involve the formation of a polyurethane structure.

Polyurethanes are widely used in foams, either flexible or rigid, but they are also employed as elastomers, fibers, surface coatings, and adhesives. Their initial exploitation was as fibre-forming materials. In 1937 Otto Bayer at I. G. Farbenindustrie in Germany hit upon the idea of using polyurethanes (with the functional groups along the polymer backbone) as rival molecules to the polyamides recently patented and commercialised in America by Du Pont under the name nylons. Bayer used di-isocyanates and diols to give linear polymers similar to the nylons, but unfortunately for him their properties were inferior to those of the nylons and polyurethanes as fibres never caught on in the way that Bayer had hoped.

As explained in Chapter 1, the urethane group is the product of the reaction of a hydroxy compound with an isocyanate group (Reaction (4.8)). This reaction occurs by step kinetics, yet is usually an addition process since no small molecule is lost as the reaction proceeds.

$$\text{ROH} \quad + \quad \text{O=C=N–R}^1 \quad \longrightarrow \quad \text{ROCONH–R}^1 \qquad (4.8)$$

The prepolymers most frequently used for the preparation of polyurethanes are either polyesters or polyethers. The polyesters are usually fully saturated and have relative molar masses in the range 1000–2000; typically such substances are viscous liquids or low melting-point solids. These polyesters are typically terminated by hydroxy-groups which act as the sites for crosslinking. Although usually linear, the polyester prepolymers may also themselves be branched. Indeed, the degree of branching determines to a large degree the properties of the finished polyurethane and this in turn determines the use to which these materials may be put. For instance, polyurethanes devived from linear polyesters may be used as elastomers; lightly branched polyesters give polyurethanes useful as flexible foams, while more heavily branched polyesters are used in polyurethanes designed for application in rigid foams.

The polyether-based polyurethanes are now of greater commercial importance then those based on polyesters. A frequently used polyether is that derived from propene oxide, as illustrated in Reaction (4.9).

$$n \quad CH_3-CH-CH_2 \longrightarrow -(CH_2-CH-O)_n- \qquad (4.9)$$

with the epoxide O over $CH-CH_2$ and CH_3 below.

The isocyanate group is reactive, a feature which leads to a large number of possible reactions when crosslinking is carried out. The essential feature of all the processes is that they involve reaction, initially at least, with an active hydrogen atom in the molecules of the co-reactant. For example, isocyanates will react with water, as illustrated in Reaction (4.10), to generate an unstable intermediate, a carbamic acid, which releases carbon dioxide to yield an amine.

$$-NCO + H_2O \longrightarrow -NH-\underset{O}{\overset{}{C}}-OH \longrightarrow -NH_2 + CO_2 \qquad (4.10)$$

Amines, too, possess active hydrogens in the sense required for reaction with an isocyanate group. Thus the products of Reaction (4.10) react further to yield substituted ureas by the process shown in Reaction (4.11). Reaction can proceed still further, since there are still active hydrogens in the urea produced in Reaction (4.11). The substance that results from the reaction between an isocyanate and a urea is called a biuret (see Reaction (4.12)).

$$-NCO + H_2N- \longrightarrow -NH-\underset{O}{\overset{}{C}}-NH- \qquad (4.11)$$

Urea

$$-NCO + -NHCONH- \longrightarrow -NHCO-NH-CONH- \qquad (4.12)$$

Biuret

Lastly, of course, the main reaction of interest is the formation of urethane groups by reaction of isocyanate groups and hydroxy-groups of the polyester or polyether. Even these reactions do not exhaust the possibilities available to the highly reactive isocyanate group. It will then go on to react with the urethane links to form a structure known as an allophanate (see Reaction (4.13)).

$$\text{—NCO} + \text{—NHCOO—} \longrightarrow \underset{\underset{\text{Allophanate}}{|}}{\text{—NHCO—N—COO—}} \tag{4.13}$$

Isocyanates are quite toxic materials and need careful handling. They affect mainly the respiratory tract causing breathing difficulties, sore throats and, in extreme cases, bronchial spasms. Once they have been allowed to react, for example to form foams, they undergo complete conversion and appear to leave no toxic residues.

Polyurethane foams are widely used. Rigid foams, for example, are used in cavity wall insulation in buildings, while flexible foams have, until recently, been used in soft furnishing for domestic use. They continue to be used in car seating. In addition to foams another major use of polyurethanes is in surface coatings. A variety of poly-urethane-based polymers, some of considerable complexity, are used for this purpose, but all share the common desirable features of toughness, flexibility, and abrasion resistance.

4.5 Epoxy Resins

Epoxy resins are those materials prepared from polymers containing at least two 1,2-epoxy-groups per molecule, generally at terminal sites of the molecule. The epoxy ring is unstable because of the high degree of strain within it and so readily undergoes reaction with a large range of substances. Most commonly employed of all the reactions undergone by the epoxy-group is addition to a proton donor species as illustrated in Reaction (4.14).

$$\underset{\text{R—CH—CH}_2}{\overset{\overset{\displaystyle O}{\diagup\!\diagdown}}{}} + \text{HX} \longrightarrow \underset{\text{R—CH—CH}_2\text{X}}{\overset{\text{OH}}{\overset{|}{}}} \tag{4.14}$$

This reaction is quite general and, since the organic group R can be aliphatic, cycloaliphatic, or aromatic, there is wide scope for variation in the composition of epoxy resins. In practice, however, the most

Figure 4.2 Diglycidyl ether.

frequently used materials are those based on bisphenol A and epichlorohydrin, which represent over 80% of commercial resins.

Epoxy resins were originally developed by Pierre Castan in the late 1930s working for the dental materials company De Trey in Zurich. They were not a success as materials for use in the mouth and the company quickly sold the patent rights to the Ciba company in Basle. After their failure as dental materials epoxy resins were developed as surface coatings and adhesives, where they showed much more promise. The surface coatings developments were due to S. O. Greenlee at Shell (USA), whose work includes the modification of epoxy resins with glycerol, epoxidation of drying oil acids, and reactions with phenolic and amino resins.

As is usually characteristic of crosslinked polymers of commercial importance, epoxy resins are prepared in two stages, with the initial reaction leading to a linear prepolymer and the subsequent reaction introducing the crosslinks between the molecules. The prepolymers from which epoxy resins are prepared are diglycidyl ethers with the structure shown in Figure 4.2.

This structure has a relative molar mass of 340; typical commerical liquid glycidyl ethers have relative molar masses in the range 340–400 and so are obviously composed largely of this substance.

Crosslinking of diglycidyl ether resins may be brought about in one of two ways, either by using catalytic quantities of curing agent or by using stoichiometric crosslinkers. Catalysts for curing epoxy resins are generally Lewis acids or bases (*i.e.* electron acceptors or donors, respectively). Consider, for example, tertiary amines which are Lewis bases and react initially with the epoxy ring as illustrated in Reaction (4.15).

$$R_3N \quad + \quad \overset{O}{\overset{\diagup\diagdown}{CH_2-CH-}} \quad \longrightarrow \quad R_3N^+-CH_2-\overset{O^-}{\underset{|}{CH}}- \qquad (4.15)$$

The ion thus produced may itself react with another epoxy-group in a process which forms the first crosslink (Reaction (4.16)). This reaction may occur at both ends of the molecule of the diglycidyl ether, so that a crosslinked structure can easily be built up from these substances. Reaction becomes complicated by the fact that the epoxy-group may also react with the hydroxy-groups that form as the epoxy ring opens up during cure. Thus the finished resin may contain a complicated array of structures within the three dimensional network.

$$-CH_2-CH- \quad + \quad CH_2-CH \longrightarrow \quad -CH_2-CH- \qquad (4.16)$$

By contrast with tertiary amines used in catalytic quantities, primary and secondary amines or acid anhydrides may be used to bring about the cure of epoxy resins by reaction in stoichiometric proportions. A typical amine curing agent used at this level is diaminodiphenylmethane (DDM), which reacts with an individual epoxy-group in the way shown in Reaction (4.17).

$$(4.17)$$

Since DDM has four active hydrogen atoms, each capable of opening up an epoxy ring, this substance is able to bring about crosslinking very readily. Unlike the reactions that occur during catalytic cure, hydroxy-groups are not able to become involved in this type of crosslinking.

Amines of this kind bring about crosslinking at room temperature and yield products having good chemical resistance. They are, however, skin irritants and require careful handling. The alternative materials for stoichiometric cure, the acid anhydrides, by contrast are not skin irritants but they will not bring about cure unless the temperature is raised. Which curing agent is chosen for use depends on a number of factors including cost, ease of handling, pot life required, and properties desired in the final resin.

5   Polymer Solutions

5.1   Introduction

In the field of polymer chemistry an understanding of polymer solutions is important. Polymers are unusual solutes in that they take
up relatively large volumes for nominally low molar concentrations
thus influencing the behaviour of polymer solutions. Such solutions
are also quite viscous, even at low molar concentration. This property
is exploited commercially in a variety of applications ranging from
paints to processed foods, where low concentrations of various polymers are used as thickening agents. Finally, an important group of
methods for characterising polymers relies on the use of polymer
solutions, including determination of relative molar mass by either
viscometry or Gel Permeation Chromatography (GPC). All of these
topics are considered in this chapter, using the thermodynamic
approaches pioneered by P. J. Flory and his co-workers in the 1940s.

5.2   Dissolution of Polymers

As we have seen previously not all polymers are capable of being
dissolved. In principle the capacity to dissolve is restricted to linear
polymers only; crosslinked polymers, while they may swell in appropriate solvents, are not soluble in the fullest sense of the word. While
individual segments of such polymers may become solvated the
crosslinks prevent solvent molecules from establishing adequate
interactions with the whole polymer, thus preventing the molecules
being carried off into solution.

The Chemistry of Polymers, 5th Edition
By John W. Nicholson
© John W. Nicholson 2017
Published by the Royal Society of Chemistry, www.rsc.org

Dissolution of polymers is a very slow process; it can take days or even weeks for particularly high relative molar mass substances. Two stages are discernible during the process of dissolution. Firstly, a swollen gel is produced by solvent molecules gradually diffusing into the polymer. Secondly, this gel gradually disintegrates as yet more solvent enters the gel and as molecules of solvated polymer gradually leave the gel and are carried out into the solution. This latter stage can be speeded up by agitation of the mixture.

Crosslinking is not the only feature that may influence solubility. Such features as crystallinity, hydrogen bonding, or the absence of chain branching may all increase the resistance of a given specimen of polymer to dissolve. Some of these features are discussed later in the chapter.

5.3   Solubility Parameters

The general principle of solubility is that like dissolves like. Hence polar polymers dissolve most readily in polar solvents, aromatic polymers in aromatic solvents, and so on. This is reflected in the thermodynamics of dissolution.

A solid will dissolve in a liquid if there is mutual compatibility; this depends on the relative magnitudes of three forces of interaction. For a polymer, p, and a solvent, s, the forces of attraction between the similar molecules are F_{pp} and F_{ss} respectively, while the force of attraction between the dissimilar molecules is F_{ps}.

In order to form a solution of polymer in solvent, F_{ps} must be greater than or equal to the forces F_{pp} and F_{ss}. If either F_{pp} or F_{ss} is greater than F_{ps} the molecules with the biggest intermolecular attraction will cohere and fail to mix with the dissimilar molecules. Under such circumstances the system will remain two-phased.

Where no specific interaction such as hydrogen-bonding can occur between the polymer and the solvent, the intermolecular attraction between the dissimilar molecules is intermediate between the intermolecular forces of the similar species, *i.e.*

$$F_{ss} < F_{ps} < F_{pp}$$

or

$$F_{pp} < F_{ps} < F_{ss}$$

If F_{ss} and F_{pp} are similar in value, then F_{ps} will be similar and the two substances will be mutually compatible and the polymer will dissolve in the solvent.

Solubility occurs where the free energy of mixing, ΔG_m, is negative. This value is related to the enthalpy of mixing, ΔH_m, and the entropy of mixing, ΔS_m, by the Gibbs equation:

$$\Delta G_m = \Delta H_m - T\Delta S_m \tag{5.1}$$

Entropy of mixing is usually (though not always) positive, hence the sign of ΔG_m is generally determined by the size and magnitude of ΔH_m. For non-polar molecules, ΔH_m is found to be positive and closely similar to the enthalpy of mixing of small molecules. In such a case, the enthalpy of mixing per unit volume can be approximated to:

$$\Delta H_m = v_s v_p (\delta_s - \delta_p)^2 \tag{5.2}$$

where v_s is the volume fraction of solvent, v_p the volume fraction of polymer. The quantity δ is known as the *solubility parameter*, with the subscripts s and p referring to solvent and polymer as before.

Using eqn (5.2), solubility parameters can be calculated for both the polymer and the solvent. Where there is no specific interaction between the polymer and the solvent, and neither has a tendency to crystallise, the polymer will generally dissolve in the solvent if $(\delta_s - \delta_p)$ is less than about 4.0; if it is much above 4.0, the polymer is insoluble in the polymer. When hydrogen-bonding occurs, however, a polymer of greatly differing δ value may dissolve in a given solvent.

To show how this system works, let us consider some of the values shown in Table 5.1. Firstly, let us use poly(ethylene) as an example. This polymer has a δ_p value of 16.2 $J\,cm^{-3}$; hexane, with a value of $\delta_s = 14.8$ $J\,cm^{-3}$ gives a $(\delta_s - \delta_p)$ of -1.4 $J\,cm^{-3}$, which being less than 4.0 indicates solubility. By contrast, methanol has a δ_s value of 29.7 $J\,cm^{-3}$, giving a $(\delta_s - \delta_p)$ of 13.5 $J\,cm^{-3}$; this indicates that poly(ethylene) is not soluble in methanol.

When using this approach to polymer solubility, we need to remember that the basis is thermodynamics. In other words, this approach gives information about the energetics of solubility, but does

Table 5.1 Solubility parameters for selected substances.

δ_s values/$J\,cm^{-3}$		δ_p values/$J\,cm^{-3}$	
n-Hexane	14.8	Poly(ethylene)	16.2
Tetrachloromethane	17.6	Poly(propylene)	16.6
Toluene	18.3	Poly(styrene)	17.6
Methanol	29.3	Poly(vinyl chloride)	19.4
Water	47.9	Nylon 6,6	27.8

not give any insight in the kinetics of the process. In order to promote rapid dissolution, it may be more helpful to employ a solvent that is less good thermodynamically, but that consists of small, compact molecules that readily diffuse into the polymer and hence dissolve the polymer more quickly.

This scheme was originally developed for non-polar systems. It can be modified to take account of polarity and of hydrogen bonding, but the resulting equations are considerably more complex.

5.4 Simple Liquid Mixtures and Raoult's Law

The free energy of dilution of a solution can be shown to be given by:

$$\Delta G_{dil,A} = kT \ln(p_A/p_A^{\circ}) \tag{5.3}$$

where p_A is the vapour pressure of substance A in the solution and p_A° is the vapour pressure of A in the pure state.

F. M. Raoult showed that for molecules of two dissimilar types, A and B, but similar size, the vapour pressure of substance A in the mixture is related to the fraction of molecules of A in the mixture, *i.e.*

$$p_A = p_A^{\circ} \frac{N_A}{N_A + N_B} = p_A^{\circ} n_A \tag{5.4}$$

where n_A is the mole fraction of A.

Substituting back into eqn (5.3), Raoult's law shows that the free energy of mixing (or dilution) may be given by:

$$\Delta G_{dil,A} = kT \ln n_A \tag{5.5}$$

Mixtures of solvent plus solute that obey Raoult's law are described as ideal. For such solutions, heat of mixing, $\Delta H = 0$. For systems of similar sized molecules where there are no strong interactions, such as hydrogen-bonding, it is found that ΔH is close to zero.

Polymer solutions always exhibit large deviations from Raoult's law, though at extreme dilutions they do approach ideality. Generally however, deviation from ideal behaviour is too great to make Raoult's law of any use for describing the thermodynamic properties of polymer solutions.

The main reason for this is that the polymer molecules are extremely large compared with those of the solvent. To take an extreme example, consider 78 g of benzene injected into a perfectly cross-linked tractor tyre. On a molar basis, this is an extremely dilute solution: one molecule of solute in one mole of solvent. Yet, because of the extreme difference in size between the two types of molecule involved, the behaviour of such a system is nothing like that of a dilute solution.

It is found, however, that even if the mole fraction is replaced with the volume fraction, correlation with experimental results is still poor. We need an alternative approach when considering the properties of polymer solutions, and such an approach will be described in the following sections of this chapter.

5.5 Entropy of Mixing

Polymers undergoing dissolution show much smaller entropies of mixing than do conventional solutes of low relative molar mass. This is a consequence of the size of the polymer molecules: when segments of a molecule are covalently bonded to each other they cannot adopt any position in the liquid, but have to stay next to each other. Hence, the possible disordering effect when such big molecules are dissolved in solvent is much less than for molecules of, say, a typical low molar mass organic substance.

To understand this in more detail, we can consider the liquid to be based on a lattice arrangement. For low molar mass solutes, each point in the lattice can be considered to be occupied by either a solvent or solute molecule. The possible arrangements of solute and solvent molecules in an extensive lattice will be very large, as shown in Figure 5.1a. According to the Boltzmann equation,

$$S = k \ln W \tag{5.6}$$

where W is the number of different arrangements available to the system and S is the entropy. Hence, where W is large, so is S.

If we now put a polymer into this lattice, we can no longer place one molecule of solute at each lattice site. Instead, we can put only one segment of the polymer molecule at any one lattice site, as shown in Figure 5.1b. When we do this we see that there are many fewer possible arrangements for the system. The value of W is thus much lower than for the low molar mass solute, hence so is S.

We assume that polymer molecules consist of a large number of chain segments of equal length, joined by flexible links. Each link then occupies one site on the lattice. The solution has to be sufficiently concentrated that the occupied lattice sites are distributed at random, rather than having them clustered together in a non-random way.

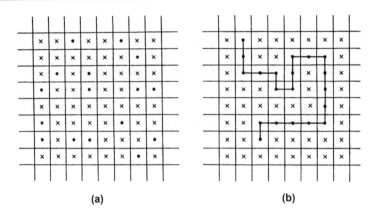

Figure 5.1 Lattice arrangements for (a) low molar mass solute in solution and (b) polymer in solution.

Using the lattice model, the approximate value of W in the Boltzmann equation can be estimated. Two separate approaches to this appeared in 1942, one by P. J. Flory, the other by M. L. Huggins, and though they differed in detail, the approaches are usually combined and known as the Flory–Huggins theory. This gives the result for entropy of mixing of follows:

$$\Delta S = - k(N_s \ln v_s + N_p \ln v_p) \tag{5.7}$$

where the subscripts denote solvent and polymer respectively, and v is the volume fraction of each component, defined as $v_s = N_s/(N_s + N_p)$ and $v_p = N_p/(N_s + xN_p)$, where x is the number of segments in the polymer.

The heat of mixing for polymer solutions, by analogy with solutions of low molar mass solutes, is given by:

$$\Delta H = \chi_s kT N_s v_s \tag{5.8}$$

The symbol χ_s stands for the interaction energy per solvent molecule divided by kT. Combining eqn (5.7) and (5.8) gives the Flory–Huggins equation for the free energy of mixing of a polymer solution:

$$\Delta G = kT(N_s \ln v_s + N_p \ln v_p + \chi_s N_s v_s) \tag{5.9}$$

The randomly occupied lattice model of a polymer solution used in the Flory–Huggins theory is not a good model of a real polymer solution, particularly at low concentration. In reality, such a solution must consist of regions of pure solvent interspersed with locally concentrated domains of solvated polymer.

A more realistic model of this solution was developed in 1950 by Flory and Krigbaum, and assumes that the polymer consists of approximately spherical clusters of segments. These clusters have a maximum density of segments at their centre and this density decreases with distance from the centre in an approximately Gaussian distribution.

As is well known from conventional physical chemistry, we can evaluate a term known as the chemical potential of a species from the variation of ΔG with changes in the amount of that species, keeping all other conditions and composition constant, *i.e.*

$$\mu_i - \mu_i^\circ = \left(\frac{f\Delta G}{fn_i}\right)_{T,P,n_j} \tag{5.10}$$

The approach of Flory and Krigbaum was to consider an excess (E) chemical potential that exists arising from the non-ideality of the polymer solution. Then:

$$(\mu_i - \mu_i^\circ)^E = -RT(1/2 - \chi)\phi_2^2 \tag{5.11}$$

This involves the Flory–Huggins parameter χ and hence assumes the same limitation as the rest of the Flory–Huggins approach, *i.e.* a 'moderately concentrated' solution. Flory and Krigbaum rewrote this equation in terms of some other parameters, *i.e.*

$$(\mu_i - \mu_i^\circ)^E = -RT(1 - \theta/T)\psi_1\phi_2^2 \tag{5.12}$$

where ψ_1 is an entropy parameter. This expression no longer assumes 'moderate concentration' but is in principle applicable to a much wider range of concentrations of polymer in solvent.

The term θ is important; it has the same units as temperature and at critical value $(\theta = T)$ causes the excess chemical potential to disappear. This point is known as the θ temperature and at it the polymer solution behaves in a thermodynamically ideal way.

5.6   Real Molecules in Dilute Solution

Two segments of a given polymer molecule cannot occupy the same space and, indeed, experience increasing repulsion as they move closer together. Hence the polymer has around it a region into which its segments cannot move or move only reluctantly, this being known as the *excluded volume*. The actual size of the excluded volume is not fixed but varies with solvent and temperature.

Typically in solution, a polymer molecule adopts a conformation in which segments are located away from the centre of the molecule in an approximately Gaussian distribution. It is perfectly possible for any given polymer molecule to adopt a very non-Gaussian conformation, for example an *'all-trans'* extended zig-zag. It is, however, not very likely. The Gaussian set of arrangements are known as *random coil* conformations.

Solvents for a particular polymer may be classified on the basis of their θ temperatures for that polymer. Solvents are described as 'good' if θ lies well below room temperature; they are described as 'poor' if θ is above room temperature.

In good solvents, a polymer becomes well solvated by solvent molecules and the conformation of its molecules expands. By contrast, in poor solvents a polymer is not well solvated, and hence adopts a relatively contracted conformation. Eventually of course, if the polymer is sufficiently 'poor' the conformation becomes completely contracted, there are no polymer–solvent interactions, and the polymer precipitates out of solution. In other words, the ultimate poor solvent is a non-solvent.

One factor which affects the extent of polymer–solvent interactions is relative molar mass of the solute. Therefore the point at which a molecule just ceases to be soluble varies with relative molar mass, which means that careful variation of the quality of the solvent can be used to fractionate a polymer into fairly narrow bands of polymer molar masses. Typically, to carry out fractionation, the quality of the solvent is reduced by adding non-solvent to a dilute solution of polymer until very slight turbidity develops. The precipitated phase is allowed to settle before removing the supernatant, after which a further small amount of non-solvent is added to the polymer solution. Turbidity develops once again, and again the precipitated phase is allowed to settle before removal of the supernatant. Using the technique polymers can be separated, albeit slowly, into fractions of fairly narrow relative molar mass.

5.7   Shapes of Polymer Molecules in Solution

In solution the molecules of a polymer undergo various segmental motions, changing rapidly from one conformation to another, so that the molecule itself effectively takes up more space than the volume of its segments alone. As we have seen, the size of the individual molecules depends on the thermodynamic quality of the solvent; in 'good' solvents chains are relatively extended, whereas in 'poor' solvents they are contracted.

The typical shape of most polymer molecules in solution is the random coil. This is due to the relative ease of rotation around the bonds of the molecule and the resulting large number of possible conformations that the molecule can adopt. We should note in passing that where rotation is relatively hindered, the polymer may not adopt a random coil conformation until higher temperatures.

Because of the random nature of the typical conformation, the size of the molecule has to be expressed in terms of statistical parameters. Two important indications of size are:

Root-mean-square end-to-end distance, $(r^2)^{1/2}$, which effectively takes account of the average distance between the first and the last segment in the macro-molecule, and is always less that the so-called *contour length* of the polymer. This latter is the actual distance from the beginning to the end of the macro-molecule travelling along the covalent bonds of the molecule's backbone.

Radius of gyration, $(s^2)^{1/2}$ which is the root-mean-square distance of the elements of the chain from its centre of gravity. This is a useful indication of size since it can readily be determined experimentally, using viscometry. For linear polymers not extended much beyond their most probable shape, these two parameters are related by the expression:

$$r^2 = 6s^2 \tag{5.13}$$

If we consider the size of a polymer molecule, assuming that it consists of a freely rotating chain, with no constraints on either angle or rotation or of which regions of space may be occupied, we arrive at the so-called 'unperturbed' dimension, written $(r)_0^{1/2}$. Such an approach fails to take account of the fact that real molecules are not completely flexible, or that the volume element occupied by one segment is 'excluded' to another segment, *i.e.* in terms of the lattice model of a polymer solution, no lattice site may be occupied twice. Real molecules are thus bigger than the unperturbed dimension, which may be expressed mathematically as:

$$(r^2)_0^{1/2} a = (r^2)^{1/2} \tag{5.14}$$

where a is called the linear expansion factor. The value of a depends on the nature of the solvent and is bigger in thermodynamically good solvents than in poor solvents.

In a sufficiently poor solvent at a given temperature, the condition where $a = 1$ can be achieved, and the chain attains its unperturbed dimensions. This turns out to be the θ temperature of Flory and Krigbaum previously described in Section 5.5 of this chapter.

Since the value of $(r^2)_0^{1/2}$ is a property of the polymer only, depending as it does only on chain geometry, it follows that the condition of the polymer at the θ temperature in different solvents is exactly the same. The polymer behaves as though it were thermodynamically ideal showing no interaction at all with the solvent.

5.8   Reptation Model of Molecular Motion

The reptation model was developed by the Nobel Prize winner Pierre de Gennes of the Sorbonne, Paris, in 1971. It has since been extended by Sir Sam Edwards and Masao Doi. The model describes the thermal motion of very long, linear polymer molecules in either concentrated solutions or melts, and their critical feature is that they are able to form *entanglements*. The term "entanglement" describes a physical interaction between polymer molecules that restrict their movement relative to other molecules.

The model envisages individual polymer molecules as moving in a wiggling, snake-like manner. De Gennes termed this movement *reptation*, a word which comes from the Latin for creeping and is also the root of the word reptile. This movement can be modelled as occurring within a narrow tube formed by all the other polymer molecules. Although each of the polymer molecules forming the tube is undergoing molecular motion, the tube remains in position for timescales that are long compared with the time it takes for the reptating molecule to pass through.

There are two major assumption involved in the model:

(i) That there is no lateral wriggling of the polymer molecule. Instead, it fits snugly inside the tube.

(ii) That there is no dynamic coupling of the motions of the chains, but each moves entirely on its own. In reality, this is not likely to be true, and motion by one polymer molecule is likely to involve other molecules being dragged along as well. However, for simplicity this kind of co-operative motion is ignored in the model.

The reptation model describes the effect of polymer chain entanglements on various processes involving molecular motion, such as diffusion and fracture. Specifically, it gives rise to two important predictions, namely:

$$D_s \propto 1/M^2$$

where D_s is the self-diffusion coefficient of the polymer and M is the molar mass (assumed to be monodisperse; see Chapter 6); and

$$K_1 \propto M^2$$

where K_1 is the fracture toughness (see Chapter 7) and, again, M is the molar mass of the polymer.

The beauty of the reptation model is that it is able to make predictions about molecular flow in both the liquid phase and at fracture by assuming that the molecules undergo the same kind of motions in each case. In both situations, the force to be overcome by the reptating polymer molecule passing through the tube is assumed by be frictional.

In a number of experimental studies, molar mass exponents close to 2 have been found for both self-diffusion and fracture toughness, though always with some deviations. Thus, using radio-labelled molecules, the diffusion coefficient for poly(styrene) in dibutyl phthalate has been found to be 2.5. Similarly, for poly(methyl methacrylate), the fracture toughness has been found to vary with $M^{2.5}$. Variations from the predicted relationship have been attributed to the polydispersity of the polymers involved. Results such as these seem to confirm the essential concepts of the reptation model, and it is now generally accepted as being a valid picture of how polymer molecules behave.

It is important to realise that the reptation model applies only to high molar mass linear polymers, *i.e.* those long enough to form proper chain entanglements. The relationship of these flow properties to molar mass with shorter chains, and hence unable to form entanglements, is different, and the proportionality with M^2 does not apply. The model also does not apply to macromolecules with more complex architectures such as branched, star or comb morphologies.

5.9 Water-soluble Polymers

A number of synthetic polymers that are widely used commercially are soluble in water. These tend to have very polar functional groups and include such polymers as poly(vinyl alcohol), poly(acrylic acid), and the modified celluloses.

Mere possession of polar functional groups is not, by itself, enough to confer water-solubility. Poly(vinyl alcohol), which is prepared by the hydrolysis of poly(vinyl ethanoate), is only soluble if a few ethanoate groups are left unreacted. Ironically the presence of these few relatively non-polar groups makes this polymer more water-soluble. The reason is that the non-polar groups interrupt the regularity of the structure in the solid polymer, thus making possible the entry of water on this material. This in turn paves the way for solvation of the polar functional groups of the individual molecules leading to dissolution. Such insolubility arises for kinetic rather than thermodynamic reasons.

In the case of water-soluble polymers, there is another factor that has to be taken into account when considering solubility, namely the possibility of hydrophobic interactions. If we consider a polymer, even one that is soluble in water, we notice that it is made up of two types of chemical species, the polar functional groups and the non-polar backbone. Typically, polymers have an organic backbone that consists of C—C chains with the majority of valence sites on the carbon atoms occupied by hydrogen atoms. In other words, this kind of polymer partially exhibits the nature of a hydrocarbon, and as such resists dissolution in water.

Hydrophobic interactions of this kind have been assumed to originate because the attempt to dissolve the hydrocarbon component causes the development of cage structures of hydrogen-bonded water molecules around the non-polar solute. This increase in the regularity of the solvent would result in an overall reduction in entropy of the system, and therefore is not favoured. Hydrophobic effects of this kind are significant in solutions of all water-soluble polymers except poly(acrylic acid) and poly(acrylamide), where large heats of solution of the polar groups swamp the effect.

The hydrophobic interaction results in the existence of a lower critical solution temperature and in the striking result that raising the temperature reduces the solubility, as can be seen in liquid–liquid phase diagrams (see Figure 5.2a). In general, the solution behaviour of water-soluble polymers represents a balance between the polar and the non-polar components of the molecules, with the result that many water-soluble polymers show closed solubility loops. In such cases, the lower temperature behaviour is due to the hydrophobic effects of the hydrocarbon backbone, while the upper temperature behaviour is due to the swamping effects of the polar (hydrophilic) functional groups.

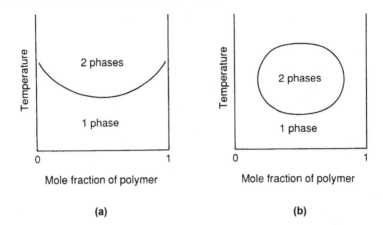

Figure 5.2 Liquid–liquid phase diagrams for (a) aqueous solutions of poly(methacrylic acid) and (b) aqueous solutions of poly(ethylene glycols).

5.10 Uses of High-viscosity Polymer Solutions

The fact that very low concentrations of polymer give highly viscous solutions is exploited commercially in a number of applications. The thickening action of polymers is often necessary for water-based substances, such as foods, toothpastes, or emulsion paints, but examples also occur of the use of polymers to thicken solvent-based products, such as paint stripper.

The main polymers used as 'thickeners' are modified celluloses and poly(acrylic acid). Several different modified celluloses are available, including methyl-, hydroxypropyl methyl-, and sodium carboxymethylcellulose and their properties vary according to the number and distribution of the substituents and according to relative molar mass of the parent cellulose. Hence a range of materials is available, some of which dissolve more readily than others, and which provide a wide spread of possible solution viscosities. Poly(acrylic acid) is also used as a thickener, and is also available in a range of relative molar masses which give rise to give solutions of different viscosities.

There are numerous applications where the development of high viscosity is necessary in a finished product. For example, thickeners, mainly based on poly(acrylic acid), are used to give 'body' to so-called emulsion paints. Emulsion paints are not formulated from true emulsions (*i.e.* stable dispersions of organic liquids in water), but are prepared from latexes, that is, dispersions of polymer in water.

Since latexes do not contain soluble polymers, they have a viscosity almost the same as pure water. As such, they would not sustain a pigment dispersion, but would allow it to settle; they would also fail to flow out adequately when painted on to a surface. Inclusion of a thickener in the formulation gives a paint in which the pigment does not settle out and which can readily be applied by brush to a surface.

Paint strippers are also formulated to have high viscosity, otherwise they run off vertical surfaces and thereby fail to penetrate or solubilise the paint to which they have been applied. Hydroxypropyl methylcellulose is the main thickener for paint strippers, which use methylene chloride (dichloromethane) as the principal component. Hydroxypropyl methylcellulose is useful for this purpose because it is soluble in the organic component but is not sensitive to the presence of any water that may also be present in the paint stripper.

Sodium carboxymethylcellulose is acceptable for use in food, and is employed in a variety of foodstuffs. It is used to prevent formation of ice crystals in ice creams; to control the consistency of cheese spreads; to stabilise the emulsions needed in salad creams; and to thicken toothpaste.

Other uses of thickening agents include pharmaceutical preparations, paper production, and oil well drilling fluids. This latter use is necessary because oil is obtained from rock that is porous. In order to remove the oil without altering the mechanical properties of the porous rock, viscous liquids ('drilling fluids') are pumped into the rock to replace the oil. Among the substances that can be used for this purpose are thickened aqueous solutions of polymers such as poly(acrylic acid) or poly(acrylonitrile).

Thus, as this short section has shown, the fact that polymer solutions are non-ideal in the sense that they do not obey Raoult's law leads to numerous important applications in the world beyond the chemical laboratory. The use of polymers as thickeners, while lacking the apparent glamour of some applications of these materials, is significant commercially and accounts for the consumption of many tonnes of polymer throughout the world each year.

5.11 Polymer Melts

Linear polymers when heated sufficiently undergo transition from solid to liquid, that is, they melt. Liquids are characterised in general by greater disorder than solids and by substantially increased molecular mobility.

To understand the nature of liquid flow in molten polymers we can again turn to the lattice model of the liquid state. Unlike the case of a polymer dissolved in solvent, for a molten polymer not all of the lattice sites are occupied. Some may be vacant, though as in the case of a solution, the remainder can be occupied by no more than one segment of the polymer chains. During molecular motion in a polymer melt the vacant sites or holes can be envisaged as jumping about and effectively swapping sites with individual polymer segments.

When a stress is applied to the bulk polymer melt, the mass flows in the direction that relieves the stress. At the molecular level, the probability of a molecular jump becomes higher in the direction of the stress than in any other direction and hence these stress-relieving motions predominate, leading to the observed pattern of flow. There is evidence that the molecular unit of flow is not the complete macromolecule but rather a segment of the molecule containing up to 50 carbon atoms. Viscous flow takes place by successive jumps of such segments until the entire macromolecule has shifted.

Polymer chains are strongly entangled in the melt but despite this they behave in a way that is thermodynamically ideal. This surprising fact was first reported by P. J. Flory in 1949, but may be readily understood. If we consider the repulsion potential, U, experienced by a monomer unit of a polymer in the melt, we can divide U into two terms, one due to repulsion by other monomer units of the same molecule, U_s, and the other due to repulsion by monomer units in different polymer molecules, U_d, *i.e.*

$$U = U_s + U_d \qquad (5.15)$$

For our arbitrary segment, at any distance $U_s + U_d$ gives the same value (*i.e.* U is constant), since the two component terms vary in opposite directions. U_s is a maximum at the position of the actual segment and reduces with increasing distance, whereas U_d is a minimum at the segment but increases with increasing distance.

The overall result is that in the melt the polymer molecules adopt Gaussian configurations and behave as thermodynamically ideal entities. This combination of ideality and chain entanglement has been confirmed by neutron scattering experiments and is well established despite the apparent paradox.

Unlike the case with dilute solutions of polymer, the variation of the melt viscosity and molar mass is far from completely understood. However, the melt viscosity, η_m, has been found to vary uniformly with number of carbon atoms in the chain above about 300–500, according to the equation:

$$\eta_m = K(DP)^{3.4}$$

where K is a constant and DP is the degree of polymerisation. This equation has been found to apply to a variety of polymers, including poly(styrene), poly(ethylene glycol), poly(butadiene), poly(ethylene) and poly(dimethoxysilane).

The practical effect shown by this equation is that polymers become more difficult to process as their molar mass increases. For example, doubling the degree of polymerisation leads to an approximately ten-fold increase in melt viscosity. Fortunately, melt viscosity decreases with increasing temperature, so that in many cases the effect of high viscosity for higher molar masses can be overcome. However, there is an upper limit at which polymers can be processed without beginning to degrade so it follows that, at some point, a polymer cannot be processed from the melt at all.

6 Methods of Determining Relative Molar Mass

6.1 Introduction

The methods by which polymers are prepared result in a mixture of molecular sizes whose properties depend on the average size of the molecules present. In principle there are a number of ways in which such an average can be calculated. The most straightforward is the simple arithmetic mean, usually called the *number average* molar mass, M_n. This is defined by the expression

$$M_n = \frac{\Sigma N_i M_i}{\Sigma N_i} \qquad (6.1)$$

where M_i is the molar mass of the molecular species i and N_i is the number of molecules of i in the sample.

Alternatively we can define the *weight-average* molar mass where we take terms in M_i^2, *i.e.*

$$M_w = \frac{\Sigma N_i M_i^2}{\Sigma N_i M_i} \qquad (6.2)$$

For a polymer consisting of molecules all of the same molar mass $M_n = M_w$, but in all other cases, M_w is greater than M_n. We can thus use the ratio of M_w to M_n as an indication of the spread of molar masses in a particular polymer sample. This ratio is called the *polydispersity* of the polymer; where $M_w : M_n = 1$ the sample is said to be *homo-* or *mono-disperse*.

The Chemistry of Polymers, 5th Edition
By John W. Nicholson
© John W. Nicholson 2017
Published by the Royal Society of Chemistry, www.rsc.org

A further but less widely used average is the *z-average* molar mass, M_z, defined as:

$$M_z = \frac{\Sigma N_i M_i^3}{\Sigma N_i M_i^2} \qquad (6.3)$$

There is a wide variety of methods, both physical and chemical, by which relative molar mass of polymers may be determined. They include end-group analysis, measurement of colligative properties, light-scattering, ultracentrifugation, and measurement of dilute solution viscosity. All techniques involve the use of polymer solutions, the majority requiring either that the data be extrapolated to infinite dilution or that the solvent be used at the θ temperature in order to attain ideal solution behaviour. Different techniques give different averages, as illustrated in Table 6.1.

In principle all methods except viscosity measurement can be used to obtain absolute values of molar mass. Viscosity methods, by contrast, do not give absolute values, but rely on prior calibration using standards of known molar mass. The relationship between polymer solution viscosity and molar mass is merely empirical but the techniques are widely used because of their simplicity. All of the absolute methods are time-consuming and laborious and are not used on a routine basis. As well as the techniques already mentioned, there is the size-exclusion method of chromatography known as Gel-Permeation Chromatography (GPC). All of these methods are discussed in detail in the sections that follow.

The preferred term throughout this book is relative molar mass, but we should note that the use of this term is not common in polymer chemistry. More often the older term molecular weight is used, both throughout the polymer industry and among academic polymer scientists. This usage extends even to the current research literature.

Table 6.1 Experimental methods for determining different types of average relative molar mass of polymers.

Average	Experimental method
M_n	GPC
	Membrane osmometry
	Vapour phase osmometry
	End group analysis
M_w	Light scattering GPC
M_z	Ultracentrifugation

6.2 Molar Masses from Colligative Properties

In physical chemistry, we apply the term *colligative* to those properties that depend upon number of molecules present. The principal colligative properties are boiling point elevation, freezing point depression, vapour pressure lowering, and osmotic pressure. All such methods require extrapolation of experimental data back to infinite dilution. This arises due to the fact that the physical properties of any solute at a reasonable concentration in a solvent are determined not by the mole fraction of solute, but by the so-called 'activity' of the solute. This takes a value less than the actual mole fraction, and is related to it by the activity coefficient:

$$a = \gamma c \tag{6.4}$$

where a is the activity, c the concentration, and γ the activity coefficient. At infinite dilution the activity coefficient, γ has a value of unity, and hence the mole fraction is the same as the activity.

For practical purposes, the colligative property that is most useful for measuring relative molar masses of polymers is osmotic pressure. As Table 6.2 shows, all other properties take such small values that their measurement is impractical.

Colligative properties measure average relative molar masses, M_n, and in the case of osmotic pressure, Π, the important relationship is:

$$\lim_{c \to 0} \frac{\Pi}{c} = \frac{RT}{M_n} \tag{6.5}$$

From this we can develop a general expression for the relationship of these parameters to concentration. Thus:

$$\frac{\Pi}{RTc} = \frac{1}{M_n}(1 + \Gamma c + g\Gamma^2 c^2 + \dots) \tag{6.6}$$

Table 6.2 Colligative properties of a solution of polymer of molar mass 20 000 at a concentration of 0.01 g cm^{-3} (from F. W. Billmeyer, *Textbook of Polymer Science*, John Wiley & Sons, New York, 1962).

Property	Value
Boiling point elevation	1.3×10^{-3} °C
Freezing point depression	2.5×10^{-3} °C
Osmotic pressure	15 cm solvent
Vapour pressure lowering	4×10 T^3 mm Hg

In eqn (6.6), Γ is a constant and g is a function that varies according to the extent of polymer–solvent interaction, and has values close to zero for poor solvents and values close to 0.25 for good solvents.

In most cases, terms in c^2 may be neglected; where this cannot be done, we may conveniently take g to be 0.25 and rewrite eqn (6.6) as:

$$\frac{\Pi}{RTc} = \frac{1}{M_n}\left(1 + \frac{\Gamma}{2}c\right)^2 \qquad (6.7)$$

As is apparent from eqn (6.6), the way to evaluate M_n by the use of colligative properties is to plot Π/c against c. The general result is a straight line with an intercept at $c = 0$ of RT/M_n.

Alternatively in thermodynamically good solvents, where terms in c^2 are significant, we see from eqn (6.7) that the appropriate plot is of $(\Pi/RTc)^{1/2}$, from which the value of M_n itself may be evaluated.

6.2.1 Vapour Phase Osmometry

This is a widely used technique based on the determination of colligative properties. Despite its name, it is not an osmotic technique at all, but is actually an indirect method of measuring vapour-pressure lowering. The parameter that is measured is the miniscule temperature difference that is obtained in an atmosphere of saturated solvent vapour between droplets of pure solvent and droplets of polymer solution each experiencing solvent evaporation and condensation. This small temperature difference is proportional to the vapour-pressure lowering of the polymer solution at equilibrium; hence, it is also proportional to the number average relative molar mass of the solute. The arrangement of the apparatus in a vapour phase osmometer is shown in Figure 6.1.

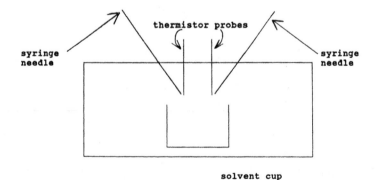

Figure 6.1 Diagram to show the essential components of a vapour phase osmometer.

The temperature differences found experimentally are less than expected theoretically because of heat losses within the apparatus. As indicated in the earlier part of this chapter, the experimental approach is to measure these temperature differences at a number of different concentrations and extrapolate to $c = 0$. The apparatus is calibrated using standard solutes of low relative molar mass, but despite this, the technique can be used on polymers up to M_n of about 40 000.

The technique is useful in that only small amounts of the sample polymer are needed, though experimentally it is time-consuming and may require great patience in use. This is because the technique does not measure equilibrium vapour-pressure lowering, but measures vapour-pressure lowering in a steady-state situation. Thus care must be taken to ensure that time of measurement and droplet size are standardised for both calibration and sample measurement.

6.3 Light Scattering

Scattering of light is a common phenomenon, since it occurs whenever light is incident upon matter. It arises because the incident beam induces vibration in nuclei and excitation of electrons when it interacts with matter. When these excited nuclei and electrons return to lower energy states, they re-emit light. Unlike the original beam, the emitted light is propagated in all directions; the wavelength, however, remains the same as in the incident beam.

Light scattering was studied by John Strutt, later Lord Rayleigh. He applied classical electromagnetic theory to the scattering of light by molecules of a gas and, in 1871, showed that one important consequence of the phenomenon of light scattering is that the sky appears blue. Rayleigh's treatment showed that, where the scattering particles are small compared with the wavelength of light, the amount of light scattered is inversely proportional to (wavelength)^4, and proportional to the number of scattering particles per unit volume.

Rayleigh's results do not apply fully to solutions. He had assumed that each particle acted as a point source independent of all others, which is equivalent to assuming that the relative positions of the particles are random. This is true in the gases with which he worked, but is not true in liquids. Hence, for solutions, the scattered light is less intense by a factor of about 50 due to interference of the light scattering from different particles.

Rayleigh showed that for light scattering, the basic relationship is

$$\frac{I}{I_0} = \frac{8\pi^4\alpha^2}{\lambda^4 r^2}(1 + \cos^2\theta) \tag{6.8}$$

where I_0 is the intensity of the incident beam, I is the intensity of the scattering radiation at a distance r from the particle and an angle of θ to the incident beam; λ is the wavelength of the radiation and α is polarisability of the particle.

In a dilute gas with molecules significantly smaller than the wavelength of light, the individual molecules act as point scatterers. For a gas consisting of N molecules in a volume of V m^3, the total scattering is N/V times the scattering from a single molecule.

The polarisability, α, of the molecule is proportional to the refractive index increment dn/dc, and to the relative molar mass of the molecule in question. The full relationship is:

$$\alpha = \frac{1}{2\pi}\frac{dn}{dc}\frac{M}{N_A} \tag{6.9}$$

Hence, the value of I/I_0 is dependent on relative molar mass of the molecules involved in the light scattering. The Rayleigh ratio, R_θ, may be defined as:

$$R_\theta = \frac{Ir^2}{I_0} \tag{6.10}$$

From this definition, eqn (6.8) may be rewritten in terms of the Rayleigh ratio:

$$R_\theta = \frac{8\pi^4\alpha^2(1 + \cos^2\theta)}{\lambda^4} \tag{6.11}$$

Taking all the constants into one super constant, K, and incorporating the result of eqn (6.9), we have

$$R_\theta = K(1 + \cos^2\theta)\,Mc \tag{6.12}$$

where c is the concentration of molecules defined as:

$$c = \frac{NM}{VN_A} \tag{6.13}$$

Eqn (6.12) is important: it shows that we can determine the relative molar mass of the molecule from the experimental measurement of the Rayleigh ratio in light scattering.

As we have seen, because of destructive interference in liquids, the intensity of the scattered light is less in liquids then in gases. The presence of solute, however, improves the situation. Fluctuations in concentration due to the presence of solute molecules cause scattering of light to be greater than that of solvent alone. In solutions, we can obtain a term that represents the Rayleigh ratio of the solute by subtracting the value of the solvent from that of the solution at a given value of r and θ.

For a polymer dissolved in solvent, where the sample is typically hetero-disperse, the expression for the Rayleigh ratio is:

$$R'_\theta = \frac{K^*(1 + \cos^2 \theta)\Sigma m_i c_i}{(1 + 2\Gamma_2 c + \ldots)} \tag{6.14}$$

where c_i is the concentration (mass per unit volume) of the molecules of molar mass m_i. Γ_2 is a complicated average. Since $\Sigma m_i c_i = c M_w$, the use of light scattering to evaluate the Rayleigh ratio leads to determination of weight average relative molar masses.

At $c = 0$, $\theta = 0$, eqn (6.14) can be rearranged to give

$$\frac{K^*(1 + \cos^2 \theta)c}{R'_\theta} = \frac{1}{M_w} \tag{6.15}$$

so that a double extrapolation to zero angle and zero concentration allows the weight average relative molar mass of a polymer to be determined. The graph that results from plotting the data obtained from light scattering experiments is called a Zimm plot, and the technique for obtaining it is referred to as Zimm's Double Extrapolation method.

In order to process the data, we carry out the following procedure. Firstly, we evaluate the term A, where

$$\frac{K^*(1 + \cos^2 \theta)c}{R'_\theta} = A \tag{6.16}$$

at varying values of scattering angle, θ. We plot the resulting values of A against $\sin^2 (\theta/2) + kc$ where k is an arbitrary spacing factor, usually 100 or 1000, chosen to separate lines of different values of c. This results is two sets of lines, one at $\theta = $ constant, the other at $c = $ constant.

Suppose we had determined a set of values of A at varying scattering angles, θ and varying concentrations of polymer in solvent, c. These are shown in Table 6.3.

Table 6.3   Values of the parameter A at varying θ and c.

	θ_1	θ_2	θ_3	θ_4	θ_5
c_1	A_{11}	A_{21}	A_{31}	A_{41}	A_{51}
c_2	A_{12}	A_{22}	A_{32}	A_{42}	A_{52}
c_3	A_{13}	A_{23}	A_{33}	A_{43}	A_{53}
c_4	A_{14}	A_{24}	A_{34}	A_{44}	A_{54}

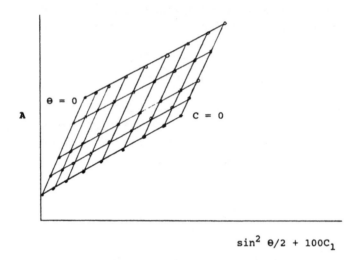

$$\sin^2 \theta/2 + 100C_1$$

Figure 6.2   A typical Zimm plot.

For $c = c_1$, the value of the term $\sin^2 (\theta/2) + 100\, c_1$ is calculated for each value of θ, the factor being an arbitrary choice. The values obtained for this term are plotted against A. The line that results is then extrapolated to $\theta = 0$, *i.e.* to $100\, c_1$. This is the uppermost line in the Zimm plot illustrated in Figure 6.2.

These calculations are repeated for c_2, c_3, and c_4 respectively. This results in a series of parallel lines inclined from the horizontal in the Zimm plot; each line is extrapolated to $100c$ to give a result for $\theta = 0$.

Once these lines are complete, another set is drawn through the data, to give a series of lines inclined from the vertical. These lines represent experimentally determined series at constant θ. These almost vertical lines are themselves extrapolated to give the $c = 0$ values.

Both extrapolated lines meet on the A axis at the same point, and this corresponds to $1/M_w$. Other solution properties of the polymer may also be determined once the Zimm plot has been prepared. Along the line of $\theta = 0$, $A = 1/M_w (1 + 2\Gamma_2\, c + \ldots)$. Hence the slope of this line is $2\Gamma_2/M_w$, from which Γ of the Flory equation may be evaluated.

Alternatively along the $c = 0$ line,

$$A = \frac{1}{M_w}\left[1 + \frac{8\pi^2 r^2}{9\lambda'^2}\sin^2(\theta/2) + \cdots\right] \tag{6.17}$$

Hence the slope of this line is $8\pi^2 r^2 / 9\,M_w\lambda'^2$, from which r^2, the square of the end-to-end distance, may be calculated. Light scattering is thus not only the primary method of determining M_w, it is also the method of choice for measuring Γ and r^2.

6.3.1 Experimental Determination

The system is set up as illustrated in Figure 6.3. In a darkened apparatus from which stray light has been eliminated an intense beam of collimated monochromatic light is passed through a cell containing the polymer solution of interest. Scattered radiation is detected using a very sensitive photo-multiplier which can be rotated through a series of accurately known angles.

Values of R'_θ are obtained partly by previous calibration using a series of standard light scatterers whose Rayleigh ratios have been precisely determined. Typical standards used in practice are poly(methyl methacrylate) blocks, colloidal silica suspensions, or tungsto-silicic acid, $H_4SiW_{12}O_{40}$.

The variable quantities in the K^* term, *i.e.* n_0^2, $(dn/dc)^2$, and λ', must be determined. Values of n_0 are available for most solvents from the literature; λ' is obtained by dividing the value of λ by the refractive index of the solution. The refractive index increment, (dn/dc), must be determined to within 10^{-5} in dn using a differential refractometer. The choice of solvent is limited: if $dn/dc = 0$, there is no scattering; if dn/dc is greater than 0.3 $cm^3\,g^{-1}$ the Rayleigh ratio is no longer proportional to $(dn/dc)^2$.

Overall, as is apparent from this description, light scattering is a difficult, time-consuming technique, despite its great importance.

Figure 6.3 Diagram of apparatus for determining molar mass by light scattering.

Despite this, the technique has been used to measure relative molar masses as low as that of sucrose and as high as those of proteins, and has been found to have a useful range for polymers of relative molar masses between ten thousand and ten million.

6.4 Viscosity Methods of Determining Relative Molar Mass

For the determination of polymer molar masses, very dilute solutions are used, typically of the order of 1% by mass. In using viscosity techniques, a number of functions are used. These are:

(i) The viscosity ratio or relative viscosity. This is defined as the ratio of the viscosity of the solution, η, to the viscosity of the pure solvent, η_0.

(ii) The viscosity number or reduced viscosity is defined by the expression

$$\frac{\eta - \eta_0}{\eta_0 c} \tag{6.18}$$

where η is the viscosity of the solution of concentration c g dm^{-3}. The viscosity ratio varies with concentration in a power series, *i.e.*

$$\frac{\eta}{\eta_0} = 1 + \eta c + k\eta^2 c^2 + \dots \tag{6.19}$$

where c is small, *i.e.* for solutions at a concentration of the order of 1% or less, terms above c^2 can be ignored as being insignificantly small. For such solutions, the viscosity number can be shown also to consist of a power series, *i.e.*

$$\frac{\eta - \eta_0}{\eta_0 c} = \eta + k\eta^2 c \tag{6.20}$$

Eqn (6.20) is known as the Huggins equation. From it, we can see that the limiting case as c tends to zero is that the term $(\eta - \eta_0)\eta_0\,c$ tends to a limiting value. This is known as the limiting viscosity number or, more rarely nowadays, the intrinsic viscosity. It is given the symbol $[\eta]$ and has units of reciprocal concentration.

We determine the value of $[\eta]$ experimentally by making measurements on a series of polymer solutions and plotting a graph of $(\eta - \eta_0)\eta_0\,c$ against c. This gives a straight line with intercept at $[\eta]$.

The value of $[\eta]$ is characteristic of the isolated polymer molecule in solution, because of the extrapolation to infinite dilution, and is a function of temperature, pressure, polymer type, solvent, and (most important of all in the present context) relative molar mass.

In practice we do not measure viscosity directly. Instead, what is measured is time of flow for solutions and pure solvent in a capillary viscometer, the so-called efflux time. If the same average hydrostatic head is used in all cases, and since for very dilute solutions the density differences between the different concentrations are negligible, then the ratio of the efflux time of the solution, t, to that of the pure solvent, t_0, may be taken as a measure of the ratio of the viscosities, *i.e.*

$$\frac{\eta}{\eta_0} = \frac{t}{t_0} \tag{6.21}$$

Similar substitutions can be made in eqn (6.20), from which we get:

$$\frac{t - t_0}{t_0 c} = \frac{\eta - \eta_0}{\eta_0 c} \tag{6.22}$$

which is the form of the Huggins equation that is used in practice.

Having obtained the value of the limiting viscosity number, we can calculate relative molar mass using the semi-empirical equation:

$$[\eta] = KM^a \tag{6.23}$$

This equation appears to have a number of names, of which the Mark–Houwink equation is the most widely used. In order to use it, the constants K and a must be known. They are independent of the value of M in most cases but they vary with solvent, polymer, and temperature of the system. They are also influenced by the detailed distribution of molecular masses, so that in principle the polydispersity of the unknown polymer should be the same as that of the specimens employed in the calibration step that was used to obtain the Mark–Houwink constants originally. In practice this point is rarely observed; polydispersities are rarely evaluated for polymers assigned values of relative molar mass on the basis of viscosity measurements. Representative values of K and a are given in Table 6.4, from which it will be seen that values of K vary widely, while a usually falls in the range 0.6–0.8 in good solvents; at the θ temperature, $a = 0.5$.

Table 6.4   Typical values of Mark–Houwink constants (from J. Brandup and E. H. Immergut, *Polymer Handbook*, Wiley, New York, 2nd edn, 1975).

Polymer	Solvent	Temperature/°C	$K \times 10^{-3}$	a
Poly(ethylene)	Decalin	70	38.7	0.67
Natural rubber	Toluene	25	50.2	0.67
Poly(styrene)	Benzene	20	6.3	0.78
Poly(styrene)	Toluene	20	4.16	0.79
PVC	THF	20	3.63	0.92

6.5   End Group Analysis

The ends of the molecules of a polymer are different from the seg-ments in the middle. For example, in polymers formed by step pro-cesses, unreacted functional groups occur. In polymers formed by chain processes, end groups occur that are either unsaturated or contain fragments of initiator. The detection and determination of these end groups can be useful in characterising macro-molecules, and may be useful in gaining understanding of the polymerisation processes themselves. However, there are difficulties with this ap-proach: unambiguous and quantitative end group analysis is only possible where these groups possess a distinctive chemical feature (such as a double bond) or contain elements that can be readily detected.

If it is possible to analyse end groups of a particular specimen of polymer, it may be possible to use the data to determine number average relative molar mass. If the molecules are branched the degree of branching can be measured from a combination of end group analysis and relative molar mass determination (determined by an alternative method).

One major drawback of end group analysis is that it rapidly becomes inaccurate as relative molar mass increases. This arises be-cause the percentage of the end groups becomes smaller and smaller, and hence more and more uncertainty attaches to the numerical values of end group content that may be obtained. To illustrate this point, let us consider a polyester with acid end groups being deter-mined by titration. Results for such titrations are shown in Table 6.5.

From Table 6.5 quantitative determination of end groups can be seen to be increasingly uncertain particularly above a value of M_n of 10 000. Nonetheless, end group analysis may be useful in certain

Table 6.5 Change in titre of 0.01 M NaOH needed to neutralise 50 mg of polyester with relative molar mass of polymer (after H. Batzer and F. Lohse, *Introduction to Macromolecular Chemistry*, John Wiley & Sons, Chichester, 2nd edn, 1979).

M_n	Volume of 0.01 M NaOH/cm^3
1000	5.0
5000	1.0
10 000	0.5
100 000	0.05

circumstances, particularly for lower molar mass polymers and oligomers, where it may be a fairly straightforward approach to obtaining useful data.

6.6   Gel Permeation Chromatography

Gel permeation chromatography (GPC) is a widely used technique for determining molar mass and molar mass distribution of polymers. In its usual form it is not an absolute method, though by making the appropriate measurements it may be made to be so.

GPC is based on the exclusion of polymer molecules above a certain size from entering particular pores located in a substance packed into a column. For this reason, it is often also referred to as *size exclusion chromatography*. The column packing may be a special porous glass or a cross-linked polystyrene gel. The mixture of polymer molecules in solution passes through the column, but depending on their size the molecules are able to permeate different volumes of pores within the packing. In the extreme case, the largest molecules may not be able to enter any pores at all, and so come through the column with the solvent front. Other molecules, which can enter a proportion of the pores, pass through more slowly. The smallest molecules of all can permeate every pore, and they thus leave the column last.

The effect is that the polymer molecules are separated into fractions. These are measured by an appropriate detector located at the end of the column, and the detector records the response as a peak on a chart. The chromatogram thus consists of a series of peaks corresponding to different elution volumes, the shortest elution volume being due to the largest molar mass polymer molecules within the sample. Details of the molar mass distribution can be determined from the size and number of the individual peaks in the

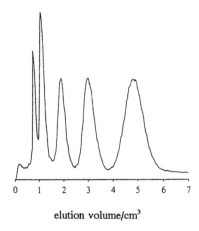

elution volume/cm³

Figure 6.4 A typical gel permeation chromatogram using monodisperse polystyrene standards.

chromatogram. An example of a gel permeation chromatogram is shown as Figure 6.4.

Gel permeation chromatograms actually give information about molecular size. For any polymer, size is determined by a number of factors. These include not only molar mass but also temperature and thermodynamic quality of the solvent. Hence the relationship between size and molar mass is unique for each particular polymer–solvent combination, and we cannot assume that because two peaks of different polymers, even in the same solvent at the same temperature, have the same elution volume their molecules have the same molar mass.

In practice, therefore, gel permeation chromatographs are usually calibrated by running monodisperse specimens of known molar mass, and determining their elution volume. From the results, a graph is plotted of $\log M$ against elution volume, which gives a straight line (see Figure 6.5). The polymer of unknown molar mass is then chromatographed, and from the elution volumes of its peaks, the values of $\log M$ and hence M itself may be determined.

Monodisperse polymers for this purpose may be obtained from commercial suppliers, but they are expensive and for many polymers not available in sufficiently narrow molar mass fractions. For this reason, the not very satisfactory procedure is sometimes adopted of calibration with polystyrene, since this polymer can be relatively easily prepared in discrete, monodisperse fractions using anionic polymerisation. The unknown polymer is then chromatographed, and its peaks related to the elution volumes for polystyrene, with the

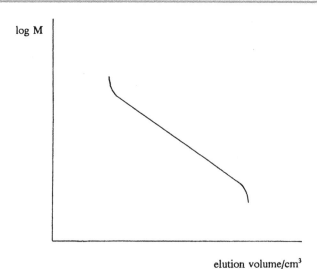

Figure 6.5 A plot of log M against elution volume showing the region of size exclusion.

resulting molar masses being described as 'polystyrene' molar masses. This is a most unsatisfactory approach.

Yet as long ago as 1966 the problem of calibration in GPC was solved. In that year, Benoit and his co-workers recognised that GPC separates on the basis of the hydrodynamic volume of the polymer molecules in solution. The intrinsic viscosity $[\eta]$ is related to the hydrodynamic volume, V_h, by the equation:

$$[\eta] = 2.5 \ A \ V_h/M \tag{6.24}$$

where A is Avogadro's number and M is molar mass. Rearranging this equation gives:

$$M[\eta] = 2.5 \ A \ V_h \tag{6.25}$$

From this it follows that plotting a graph of log $M[\eta]$ against elution volume should give a straight line for all polymer–solvent combinations. This was shown by Benoit *et al.* to be true for a large number of systems and is referred to as the *universal calibration curve*. It is illustrated in Figure 6.6.

In reality, the universal calibration curve is not quite universal. Certain exceptions exist, such as polymers with very rigid backbones. However, the equation does apply to a very large number of systems, and this means that it is possible to calibrate the chromatograph with one set of readily available monodisperse standards and to use the results to determine the real molar mass of a different polymer.

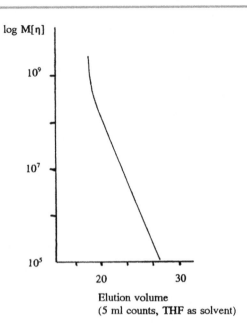

Figure 6.6   The universal calibration plot.
(As originally published by Z. Grubisic, P. Rempp and H. Benoit, *Polym. Lett.*, 1967, **5**, 753).

In order to do this, experimental determinations of the intrinsic viscosities of both the standards and the fractions from the unknown polymer are required. It is possible to obtain commercial gel permeation chromatographs that will do this routinely, and hence to exploit the concept of universal calibration. Care must be taken, though, to ensure that the separation of the polymer molecules occurs purely as a result of size exclusion. If there are any other specific interactions, *e.g.* hydrogen bonding, between the polymer and the column packing, such as may occur with water-soluble polymers, Benoit's approach does not work and the universal calibration plot is not valid.

6.7   MALDI Mass Spectrometry

In the past 20 years or so, the technique of matrix-assisted laser desorption/ionization (MALDI) mass spectrometry has been developed and increasingly applied to polymers. In MALDI, high molar mass ions are generated in the gas phase by irradiating a solid mixture of polymer distributed within a suitable matrix substance.

Irradiation is carried out with a pulsed laser beam operating at pulse times of nanoseconds in the UV region.

Samples for MALDI employ very low ratios of polymer to matrix, typically around 0.0%. As well as the matrix substance, a solvent such as acetone, water or THF is employed to dissolve the mixture. A small volume of solution, typically a few microlitres, is deposited onto a substrate, the solvent evaporated and the solid residue placed in a mass spectrometer.

Matrix substance for MALDI must be chosen carefully to ensure that the desorption and subsequent ionization steps occur effectively. In particular they should absorb strongly at the wavelength of the laser beam, and then transfer energy to the polymer sample.

Irradiation by the laser pulse causes the matrix molecule to absorb energy. The low concentration of polymer ensures that almost all of the laser energy is absorbed by the matrix substance. Once this absorption has occurred the energy can be transferred to the polymer component, causing rapid vaporization. A typical pulse removes a very small amount of material, and the dense gas formed is expanded supersonically into the vacuum within the mass spectrometer, from where it proceeds to analysis. This latter step is generally carried out in time-of-flight (ToF) mode, where molecules are accelerated towards the detector and arrive in inverse proportion to the mass, *i.e.* lighter ones arriving faster than heavier ones.

Typical MALDI matrix substances are shown in Table 6.6. These substances have to be chosen for specific polymers by a process of trial-and-error, and these substances are often used in conjunction with salts, such as LiCl, NaCl and KCl, or silver trifluoroacetate. These salts are necessary for the polymers to form positively charged quasi-molecular ions that the mass spectrometry technique is able to detect.

A typical MALDI spectrum is shown in Figure 6.7. MALDI is rapid, a typical spectrum for a polymer taking about 20 minutes to obtain. The technique can be applied to a wide range of polymers, including water-soluble polymers such as poly(acrylic acid) and poly(ethylene glycol) and both polar and non-polar organic polymers, such as poly(styrene) and poly(methyl methacrylate). Some polymers are not

Table 6.6 Frequently encountered MALDI matrix substances.

1,8,9-Anthracenetriol
2,5 Dihydroxybenzoic acid
trans-3-Indole acrylic acid
2-(4-Hydroxyphenylazo) benzoic acid

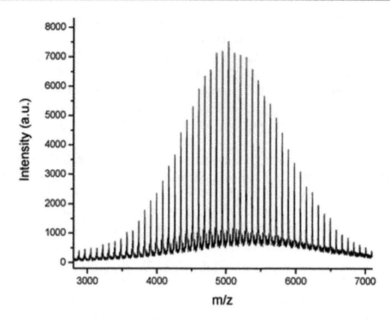

Figure 6.7   Example of a MALDI mass spectrum of a polymer sample.

easily analysed by **MALDI**, though, because of difficulties in obtaining compatible matrix substances.

Because MALDI determines the molar masses of the individual polymer molecules directly, the number average and weight average molar masses can be calculated directly using the appropriate formulae (eqn (6.1) and (6.2)). Molar masses must not be too high, due the requirement to vaporise the molecules, and values that can be determined reliably typically lie in the range up to 20 000. As a result, MALDI is often considered to be complementary to GPC, because is effective for much greater molar masses.

7 Mechanical Properties of Polymers

7.1 Introduction

The majority of polymers of commercial or technical importance are organic in nature. Yet unlike the low molecular weight compounds of conventional organic chemistry, they are not simply liquids or relatively low melting point crystalline solids. Instead they present themselves in a variety of physical forms, including liquids, rubbers, and brittle glasses or as relatively soft and flexible solids. The physical properties of these materials can be explained, broadly at least, by their underlying molecular structure. Polymers that are crosslinked into three-dimensional networks may be very brittle; uncrosslinked materials are much less so. Crystallinity of certain regions of the polymer may impart rigidity, as, if the polymer is below its glass transition temperature, may simple intermolecular forces between individual macromolecules. In addition there may be a marked anisotropy of behaviour, depending on the precise direction and orientation of the polymer molecules. The study of all these properties rightly belongs to the wider subject of materials science. However some discussion of them is appropriate in a book of this type, in order to give an appreciation of the role that chemistry has in determining the mechanical behaviour of polymeric substances.

The Chemistry of Polymers, 5th Edition
By John W. Nicholson
© John W. Nicholson 2017
Published by the Royal Society of Chemistry, www.rsc.org

7.2   Stress, Strain, and Young's Modulus

When we consider the mechanical properties of polymeric materials, and in particular when we design methods of testing them, the parameters most generally considered are stress, strain, and Young's modulus. Stress is defined as the force applied per unit cross sectional area, and has the basic dimensions of N m^{-2} in SI units. These units are alternatively combined into the derived unit of Pascals (abbreviated Pa). In practice they are extremely small, so that real materials need to be tested with a very large number of Pa in order to obtain realistic measurements of their properties. As a result, the more practical units of MPa (*i.e.* 10^6 Pa) are employed instead.

Strain is a dimensionless quantity, defined as increase in length of the specimen per unit original length. It represents the response of the material to the stress applied to it.

The ratio of stress to strain is known as Young's modulus. This parameter also has dimensions of force per unit area, but is a characteristic of the material, not merely a value imposed on it by a specific set of test conditions, as is the case for stress itself. Materials for which the evaluation of Young's modulus is particularly appropriate are those which most closely approximate to an ideal elastic solid. These materials obey Hooke's law, and for such materials the assumption that strain occurs immediately on applying the stress is essentially correct. And for such a solid, releasing the stress causes an immediate return to the original dimensions. This model of mechanical behaviour is acceptable (though not *strictly* accurate) for metals well below their melting points; it is not adequate when studying the behaviour of polymeric materials, as will be described later in this chapter.

7.3   Brittle and Tough Fracture

Probably the most basic information about a material that we need to know is its strength. But what precisely do we mean by the term 'strength'? In practice, the answer that engineers give to this question is to define the strength of a material as the force experienced at the point where it fractures. This topic is one that is complex from a theoretical point of view, largely because fracture is a point of discontinuity, and hence cannot readily be interpreted in terms of events leading up to it.

Two types of fracture are usually distinguished:

i. brittle fracture, in which the material experiences a sudden rupture with minimal deformation, and
ii. tough fracture, in which there is a substantial deformation of the specimen in the form of necking and narrowing, prior to breaking.

Whilst these extremes of behaviour can be readily distinguished, there are transitional types of fracture behaviour that lie between them, and in these cases, judgement about fracture mode can be difficult.

Strictly the terms 'brittle' and 'tough' fracture can only be applied to failure under carefully specified test conditions. That is to say that the statement that a glassy polymer, such as poly(methyl methacrylate), undergoes brittle fracture at ambient temperatures needs qualifying; test conditions must be stated. These are usually that the material has been formed into a dumbbell shaped specimen, and has been subjected to increasing tension by being drawn apart between two jaws of an appropriate machine separating at a constant rate.

The phenomena of brittle and tough fracture give rise to fairly characteristic stress–strain curves. Brittle fracture in materials leads to the kind of behaviour illustrated in Figure 7.1; fairly uniform extension is observed with increasing stress, there is minimal yield, and then fracture occurs close to the maximum on this graph.

Tough fracture, which is alternatively known as ductile fracture, by contrast, gives the type of behaviour illustrated in Figure 7.2. After the maximum in the stress–strain plot has been reached, there is a substantial amount of yielding, before the sample eventually breaks.

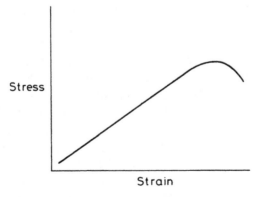

Figure 7.1 Stress–strain plot for brittle fracture.

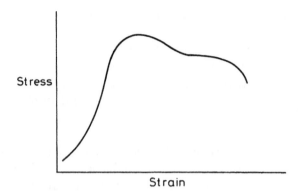

Figure 7.2   Stress–strain plot for tough fracture.

The large region of yield in materials that fail by tough fracture arises as the molecules of the polymer rearrange themselves in response to the applied stress. This is different from the mechanism of yield in metals, where planes of metal atoms slide over one another. In polymers, the molecular movement is akin to the movement in the liquid phase, and hence there is a viscous component to this rearrangement. This behaviour, which gives rise to the phenomenon of viscoelasticity, will be considered later in the chapter.

As was stated earlier, brittle and tough fracture represent the extremes of behaviour, and for any one polymer there is a transition between the two types of fracture mode. For example, in the well-known demonstration in which rubber is immersed in liquid nitrogen, the specimen undergoes a rapid conversion from a tough material into a brittle one. To explain this observation, we can regard a material as having two strengths, brittle σ_b and yield σ_y. The material will then fail by whichever of these is the smaller. Hence, a material with a lower σ_y than σ_b will fail through a process that involves significant yielding prior to fracture.

Both of these individual strengths increase with increase in the rate at which strain is applied. Yield strength has been found to be more sensitive to change in strain rate than brittle strength, and thus increases more quickly. For a material showing tough fracture at a low strain rate, this means that there comes a point at which σ_y and σ_b are equal, and this corresponds to the tough–brittle transition. Further increase in the strain rate causes σ_y to increase to a value greater than σ_b, and the sample then fails by brittle fracture. In the same way, temperature also affects the tough–brittle transition, lower temperatures tending to cause a reduction in the brittle strength so that it is lower than the yield strength. This leads to the observation that cooling makes a material more likely to be brittle, as it did for the rubber specimen mentioned earlier.

The fracture strength and mode of fracture of a material have been found to be related to a number of characteristics of the polymer molecules of which it is made up. These include, among others, constitution, molar mass, polydispersity, crystallinity, and degree of crosslinking. Other factors which also affect fracture strength and mode of fracture are temperature, strain rate, and geometry of the specimen, all of which are decided upon prior to testing the material.

The mode of fracture itself is a reflection of the position of the brittle–tough transition, which in turn reflects the relative magnitudes of σ_y and σ_b. For example, crosslinking affects yield strength more than it affects brittle strength, increasing it relative to the latter, and hence making the specimen more likely to fail by brittle fracture. Plasticisation using low molar mass additives, by contrast, lowers yield strength relative to brittle strength, thus making the specimen more likely to fail by yielding.

7.4 Types of Strength

There is no single answer to the question 'how strong is this material?' The answer depends on which particular test of strength has been applied. For example, has the material been subject to an impact, or has it been crushed? Has it been twisted or has it been pulled apart? The values for strength from these various types of test are not the same. Some materials, such as building stone and cast iron, are strong in compression (*i.e.* when crushed), but weak in tension (*i.e.* when pulled apart); other materials, such as wrought iron, are weak in compression but strong in tension. An interesting consequence of these facts was seen as the Industrial Revolution proceeded, and new engineering materials became available. This eventually led to changes in design of large-scale objects. When the first iron bridge was constructed in late 1780, at what was to become known as Ironbridge in Shropshire, the material available was cast iron, and the design of the bridge reflects the fact. It is built just as a stone bridge would have been, with all the load-bearing parts in compression. When, later, wrought iron became available, engineers were free to use other, more elegant designs, such as that which Telford used for his suspension bridge at the Menai Straits in North Wales; this bridge has the load borne by parts in tension.

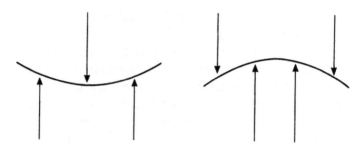

Figure 7.3   Three- and four-point loading for flexural strength estimation.

Tensile strength and compressive strength, then, are not the same, and neither can be used as the sole criterion of strength. Moreover, there are other measurements of strength in addition to these two, as indicated above. An additional possibility is flexural strength measurement, in which a bar of material, having either circular or rectangular cross-section, is bent by three or four point loading until it fractures (see Figure 7.3).

The stress systems in such tests are complex, and not easily related to fundamental properties. But the results are relevant to the performance of materials in service, and for that reason, flexural tests are frequently used in engineering practice.

Certain materials may be weak in shear, and for these it is appropriate to measure their strength by torsion tests. For such tests, the material is fabricated into a rod of circular cross-section, and twisted about its longitudinal axis. The angle of twist is proportional to the shearing strain, γ, and the applied force or couple is proportional to the shearing stress, τ, their ratio being defined as the shear stress, G, *i.e.*

$$G = \tau/\gamma \tag{7.1}$$

In addition to the strength tests already mentioned, there are other tests which can be applied and which at first sight give an indication of strength. These include hardness and fatigue tests. In reality all of these tests measure different physical attributes of the material and for a truly comprehensive picture of the behaviour of any material, results from all of these types of test should be known.

7.5   The Influence of Surfaces

The state of the surface of a brittle solid has been found to exert a considerable influence on the mechanical behaviour observed; it is at least as important as the underlying molecular constitution in this regard. The presence of microscopic scratches, voids, or other imperfections will seriously weaken the tensile strength of specimens of glassy polymer, such as poly(methyl methacrylate) at ambient temperatures.

In the 1920s, A. A. Griffith, working at the Royal Aircraft Establishment, Farnborough, turned his attention to the question of why brittle materials, such as glass (which may be regarded as a wholly inorganic polymeric material) had such low tensile strength. From a simple consideration of the strengths of the chemical bonds involved, Griffith calculated that glass ought to have a strength of about 1000 MPa, roughly the same as steel. But when he measured it, he found values of only about 50 MPa.

To find out why this was, Griffith carried out his now-famous series of experiments using freshly drawn glass threads. In this way, he found that as the fibres became thinner, so they became stronger, with the thinnest of all beginning to approach the theoretical strength predicted on the basis of the strength of the chemical bonds within the glass. These results led Griffith to the conclusion that the thicker threads of glass were covered with very fine surface cracks, introduced as the glass cooled, and that these had a weakening effect. There were fewer of these micro-cracks in the thin threads and almost none in the thinnest of all, essentially because there was no room for them.

Griffith's work led to the introduction of the equation that bears his name, and defines the relationship between breaking stress, σ_b, and crack length. According to this equation, for a brittle material containing a crack of length $2c$, and stress perpendicular to the crack, the breaking stress is given by

$$\sigma_b = \left(\frac{2\gamma E}{\pi c}\right)^{1/2} \tag{7.2}$$

where γ is the surface energy per unit length of the newly forming surfaces as the crack propagates, and E is Young's modulus. Using this equation, Griffith was able to calculate the size of the flaws in the surfaces of the glass samples he used, and found them to be of the order of 10^{-6} m in size.

The effect of these small cracks is to concentrate the stress at localised points within the specimen. Figure 7.4 illustrates how this happens, using lines to indicate the stress distribution in the sample. For the unnotched specimen, (a), the stress is uniformly distributed

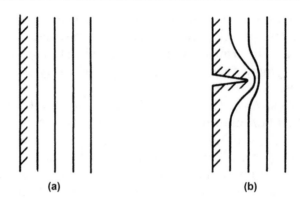

Figure 7.4 Stress concentration at surface imperfections.

throughout the material. However, for the notched specimen, (b), the lines of stress can be seen to converge at the notch tip, this giving a local stress greater than the apparent applied stress. When this happens, the breaking stress, σ_b, will occur in the material at an actual stress somewhat less than this. As a result, the material as a whole is weaker than predicted on the basis of is chemical composition.

The Griffith crack equation has been shown to apply, albeit with some scatter of results, to the brittle polymeric materials poly(methyl methacrylate) and poly(styrene) when cracks of controlled size have been introduced deliberately into the specimens. Such experiments give values of surface energy that are very large, typically 10^2–10^3 J m^{-2}, which is about 100 times greater than the theoretical value calculated from the energy of the chemical bonds involved. This value of γ thus seems to be made up of two terms, *i.e.*

$$\gamma = \gamma_E + \gamma_v \qquad (7.3)$$

where γ_E is the 'true' surface energy and γ_v represents work done at the fracture tip as polymer molecules undergo viscous flow to align themselves in response to the applied stress. Given the fact that the order of magnitude of γ is so much greater than γ_E, it follows that γ_v is much the larger term; in other words, there is substantial plastic flow ahead of the crack tip. This behaviour is more characteristic of metals than typical inorganic glasses, and indicates that the term 'glassy' for polymers such as poly(methyl methacrylate) may not be wholly appropriate, at least as far as fracture mechanisms are concerned.

As a result of this additional work to create a surface over and above the thermodynamic surface energy assumed by Griffith, we need to rewrite his equation for the relationship between breaking stress and crack length. We modify eqn (7.2) to:

$$\sigma_b = \left(\frac{K_1}{\pi c}\right)^{1/2} \tag{7.4}$$

where the term K_1 represents the *fracture toughness*. As before, eqn (7.4) is concerned with the growth of pre-existing cracks. The fracture toughness varies with the toughness; those materials in which cracks are reluctant to grow we would recognise as tough, and for them values of K_1 are large.

7.6 Viscoelasticity

No material is perfectly elastic in the sense of strictly obeying Hooke's law. Polymers, particularly when above their glass transition temperature, are certainly not. For these macromolecular materials there is an element of flow in their response to an applied stress, and the extent of this flow varies with time. Such behaviour, which may be considered to be a hybrid of perfectly elastic response and truly viscous flow, is known as viscoelasticity.

In order to model viscoelasticity mathematically, a material can be considered as though it were made up of springs, which obey Hooke's law, and dashpots filled with a perfectly Newtonian liquid. Newtonian liquids are those which deform at a rate proportional to the applied stress and inversely proportional to the viscosity, η, of the liquid. There are then a number of ways of arranging these springs and dashpots and hence of altering the mathematical relationship between the elastic and the viscous elements of the stress–strain relationship. The simplest combination, which is the only one that will be considered here, has the spring and dashpot in parallel, as illustrated in Figure 7.5. This parallel coupling is known as the Kelvin or Voigt arrangement. The total strain in such a system is equal to the strain of either component, its magnitude being determined by whichever of the components offers more resistance to the stress. For a constant applied stress, the initial component to take up the stress is the liquid in the dashpot. It deforms slightly, causing the spring in turn to take up a small proportion of the overall stress that has been applied. As time progresses, this process continues until all of the stress is carried by the spring and none at all is carried by the liquid in the dashpot.

Figure 7.5 Parallel arrangement of spring and dashpot as used to describe viscoelasticity.

A real material whose behaviour can be modelled in this way initially undergoes irreversible deformation as the stress is applied. This eventually ceases, and the material then behaves effectively as an elastic solid. Release of the stress will cause a rapid return to a less strained state, corresponding to the 'spring' component of the response, but part of the deformation, arising due to viscous flow in the 'dashpot' will not disappear.

The relationships between stress, σ, and strain, ε, are as follows:

For the spring, Hooke's law is obeyed and

$$\sigma_s = E\varepsilon \tag{7.5}$$

where E is Young's modulus.

For the dashpot,

$$\sigma_d = \eta \frac{d\varepsilon}{dt} \tag{7.6}$$

In the parallel coupling of the Kelvin or Voigt model, the applied stress, σ_0, is simply the sum of the stresses of the individual components:

$$\sigma_0 = \sigma_s + \sigma_d$$
$$= E\varepsilon + \eta \frac{d\varepsilon}{dt} \tag{7.7}$$

This is the simplest way of applying the spring and dashpot model, but there are others of increasing complexity. For example, the Maxwell model considers the spring and dashpot to be in series, while the so-called standard linear solid has both parallel and series arrangements. While all of these approaches are mathematically useful, they do not have an underlying physical basis; in reality there are no springs and no dashpots. However, what actually does happen on the molecular level is complicated. In the unstressed state, the polymer molecules are undergoing certain movements anyway, simply as a result of possessing thermal energy. When stress is applied, the movements of the molecules are such that the material deforms in the manner described above. Some of the response is rapid, almost instantaneous, and is modelled as the spring. Other changes, involving more long-range interactions or extensive adjustments in polymer configuration or orientation, are slower, and give rise to the viscous element.

7.7 Creep and Stress Relaxation

There are two further related sets of tests that can be used to give information on the mechanical properties of viscoelastic polymers, namely creep and stress relaxation. In a creep test, a constant load is applied to the specimen and the elongation is measured as a function of time. In a stress relaxation test, the specimen is strained quickly to a fixed amount and the stress needed to maintain this strain is also measured as a function of time.

In creep tests, the parameter of interest is the creep compliance, J, defined as the ratio of the creep strain to the applied stress, *i.e.*

$$J = \varepsilon/\sigma \qquad (7.8)$$

Polymers which creep readily have large values of J; polymers which hardly creep at all have small values. For viscoelastic polymers below their glass transition temperature, there is a characteristic creep curve, as illustrated in Figure 7.6.

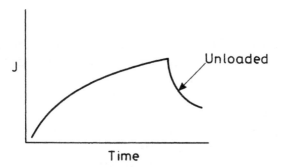

Figure 7.6 Typical creep curve for a viscoelastic polymer.

Immediately the load is applied, the specimen elongates corresponding to an instantaneous elastic modulus. This is followed by a relatively fast rate of creep, which gradually decreases to a smaller constant creep rate. Typically this region of constant creep in thermoplastics essentially corresponds to viscous flow. In terms of the spring and dashpot model, the retardation is dominated by the viscous liquid in the dashpot. As before,

$$\sigma = \eta \frac{d\varepsilon}{dt} \tag{7.9}$$

Since the stress is constant, it follows that so also is the creep rate. The creep compliance at time t, J_t, can be considered to consist of three terms, an instantaneous compliance, J_0, a term covering a variety of retardation processes, $\psi(t)$, and a viscous term, t/η. These are related by:

$$J_t = J_0 + \psi(t) + t/\eta \tag{7.10}$$

The last term represents irrecoverable flow which occurs in these polymers such that there is a permanent deformation which remains in the specimen after the load is removed.

There are varieties of creep behaviour, depending on the nature of the polymer. Amorphous polymers well below their glass transition temperature do not creep much at all. As the temperature is raised, past T_g, and into the region of rubbery behaviour, so creep rate increases. In the latter state, creep rate is affected by molar mass of the polymer molecules, higher molar mass polymers flowing less readily than those of lower molar mass. Slightly crystalline polymers do not creep much at all until well above their glass transition temperatures. The reason for this is that the crystalline regions behave like crosslinks and hence inhibit the movement of the polymer molecules.

Stress relaxation tests are alternative ways of measuring the same basic phenomenon in viscoelastic polymers as creep tests, *i.e.* the time-dependent nature of their response to an applied stress. As such, they have also been of value in understanding the behaviour of these materials. The essence of stress relaxation tests is that strain increases with time for a given stress, so that if stress is decreased with time in a controlled manner ('relaxed'), a state of constant strain can be maintained. From such experiments, a stress relaxation modulus, $G(t)$ at time t s after setting the strain can be determined, defined as $G(t) = \sigma(t)/\varepsilon$.

7.8 Cold Drawing

Thermoplastic polymers subjected to a continuous stress above the yield point experience the phenomenon of cold-drawing. At the yield point, the polymer forms a neck at a particular zone of the specimen. As the polymer is elongated further, so this neck region grows, as illustrated in Figure 7.7.

As deformation continues, so the neck region becomes longer until the entire specimen has been drawn out into the new shape with a cross-sectional area that of the original neck zone. Once the polymer is fully cold drawn, it is stronger than during neck propagation, and hence there is a final upswing in the stress–strain curve (see Figure 7.8). The reason for this increase in strength is that as cold drawing takes place, so polymer molecules align themselves with the direction of stress.

Figure 7.7 Successive stages in cold-drawing of a polymer.

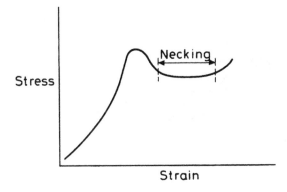

Figure 7.8 Stress–strain curve for polymer during cold-drawing.

Amorphous polymers may be cold drawn only below their glass transition temperatures; above this temperature, they stretch but without forming a well-defined neck region. Crystalline polymers, by contrast, can be cold drawn at all temperatures up to almost their melting points.

7.9 Dynamic Tests

Dynamic tests are extremely useful for evaluating the stress relaxation and creep behaviour of polymeric materials. Broadly, such tests are carried out by setting a specimen of the material into vibration, and then varying the frequency that is applied. These variations in the applied frequency lead to changes in the frequency of vibration of the specimen, such changes being related to the nature and extent of the relaxation processes occurring in the polymer. A wide range of frequencies, varying from 0.01 to 300 Hz, has been used in this kind of experiment, and in this way a substantial amount of detail has been discovered about the mechanical properties of viscoelastic polymers.

The general mode of operation in dynamic tests is to vary the stress sinusoidally with time. A viscoelastic solid in which the viscous element is that of a Newtonian liquid (as defined earlier) responds with a sinusoidal strain of identical oscillation frequency. However, because of the time-dependent relaxation processes taking place within the material, the strain lags behind the stress, as illustrated in Figure 7.9.

For a frequency of oscillation of $\omega/2\pi$ Hz, the stress σ at any given time, t, is given by:

$$\sigma = \sigma_0 \sin \omega t, \qquad (7.11)$$

where σ_0 is the maximum stress.

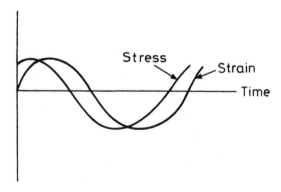

Figure 7.9 Lag of strain behind stress in dynamic testing.

The corresponding strain is given by:

$$\varepsilon = \varepsilon_0 \sin(\omega t - \delta) \tag{7.12}$$

where δ is the phase angle, and represents the amount that the strain lags behind the stress.

The stress can be resolved into two parts, one in phase with the strain and the other 90° out of phase with the strain.

Assuming these conditions apply to a torsion experiment, it also becomes possible to define two shear moduli, G_1 and G_2. The first of these represents that part of the stress that is in phase with the strain, divided by the strain,

$$G_1 = \sigma_1/\varepsilon_0 \tag{7.13}$$

The second shear modulus, G_2, is the peak stress 90° out of phase with the strain, divided by the peak strain,

$$G_2 = \sigma_2/\varepsilon_0$$
$$= \frac{\sigma_1 \tan \delta}{\varepsilon_0} \tag{7.14}$$

The value of this latter parameter is proportional to the energy dissipated as heat per cycle, and is known as the loss modulus. The former quantity, G_1, is proportional to the recoverable energy, and is called the storage modulus. The two are combined to form the complex modulus, G^*, related by the equation

$$G^* = G_1 + iG_2 \tag{7.15}$$

where

$$i = \sqrt{-1}.$$

The relative magnitudes of these two moduli, G_1 and G_2, vary according to the state of the polymeric material. In the glassy state, where good elasticity is shown, G_1 is high; in the rubbery state, where there is a greater contribution from the viscous element, G_1 is low.

In the glassy state, polymer molecules are highly restricted in their motion (hence the almost perfect elasticity of such materials) and any imposed vibration experiences little damping. Essentially no strain energy is lost as heat. At the glass transition region, parts of the polymer molecules become free to move, though with some delay in their response to an applied stress. This results in high damping, with much strain energy lost as heat.

Once beyond the transition region, and well into the rubbery state, all molecular segments are free to move. There is no inhibition to movement, hence little time delay in molecular response to an applied stress. In this situation, strain energy is not dissipated as heat, but appears as a decrease in entropy. Unlike dissipated heat, such a decrease in entropy can be recovered and, as a result, damping decreases again. Thus, consideration of dynamic experiments leads to a further criterion of the glass transition being identified; it is the point of maximum damping in appropriate dynamic tests.

7.10 Time/Temperature Relationship

In order to have enough data to characterise the viscoelastic behaviour of an amorphous polymer fully, it is necessary to collect data on relaxations over about ten to fifteen decades of time. In practice, this is extremely tedious, and may not even be possible. To overcome the problem, use is made of the fact that, for most amorphous polymers, a deformation for a short period of time at one temperature is equivalent to a longer period at a lower temperature. Essentially identical relaxation processes occur, differing only in how long they take. Thus, it becomes possible to build up a 'master-curve' of relaxation modulus, $G(t)$, against time at a single arbitrary temperature by suitable processing of data obtained at a variety of temperatures. A typical master curve is shown in Figure 7.10.

In order to produce the master curve illustrated, each section would have been completed in the time range 10^{-1}–10^4 s, but at different temperatures. Combining the sections then produces the overall master curve.

In order to convert data obtained at a given experimental temperature, T, to the reference temperature, T_0, a shift factor, a_T is used, and is defined as

$$a_T = t_T/t_{T_0} \qquad\qquad (7.16)$$

Figure 7.10   Stress relaxation master curve at a given temprature.

where t_T is the relaxation time at temperature T, and t_{T_0}, is the relaxation time at the reference temperature, T_0.

The value of a_T itself is obtained by using the so-called WLF eqn (7.18), first proposed by Williams, Landel, and Ferry in 1955.

$$\log a_t = \frac{-17.44(T-T_0)}{51.6 - (T-T_0)} \tag{7.17}$$

The glass transition temperature can be chosen as the reference temperature, though this was not recommended by Williams, Landel, and Ferry, who preferred to use a temperature slightly above T_g. In order to determine relaxation times, and hence a_t, use can be made of dynamic mechanical, stress relaxation, or viscosity measurements.

The WLF equation can be widely applied, and demonstrates the equivalence of time and temperature, the so-called time–temperature superposition principle, on the mechanical relaxations of an amorphous polymer. The equation holds up to about $100°$ above the glass transition temperature, but after that begins to break down.

7.11 Rubberlike Elasticity

The elasticity of certain materials including vulcanized natural rubber at ambient temperatures is well known. When such materials are subjected to a stress, they experience a rapid and clearly defined strain, and this strain disappears immediately the stress is removed. Moreover, they undergo very large deformations under an applied stress; vulcanized natural rubber, for example, can be extended to 1500% of its original length. Such behaviour is the essence of rubberlike elasticity. Unlike viscoelastic materials, there is virtually no creep. In terms of the spring and dashpot model described earlier, the material behaves almost entirely as a spring, with negligible contribution from viscous flow in the dashpot.

In reality the ideal elastic rubber does not exist. Real rubbery materials do have a small element of viscosity about their mechanical behaviour, even though their behaviour is dominated by the elastic element. Even so, real rubbers only demonstrate essentially elastic behaviour, *i.e.* instantaneous strain proportional to the applied stress, at small strains.

Such rubberlike elasticity is exhibited only by macromolecular materials, generally those in which there is a small amount of crosslinking. Raw natural rubber, for example, is not elastic until it has been lightly crosslinked. This is achieved by incorporating a small amount of sulfur, of the order of a few percent by weight, together with an appropriate accelerator, followed by heating. This treatment with sulfur, so-called vulcanization, links the macromolecules at particular points to form a light network structure which gives the material the elasticity generally regarded as characteristic of rubber. A number of other polymers can be prepared which behave similarly, these being known as synthetic rubbers or elastomers. The latter name is to be preferred, since it leaves no room for doubt about the mechanical nature of the material.

In the unstressed state the molecules of an elastomer adopt a more-or-less randomly coiled configuration. When the elastomer is subjected to stress the bulk material experiences a significant deformation, as the macromolecules adopt an extended configuration. When the stress is removed, the molecules revert to their equilibrium configurations, as before, and the material returns to its undeformed dimensions.

In a non-rigorous way we can use thermodynamics to explain the phenomenon of the elasticity of vulcanized natural rubber. Measuring enthalpy (*i.e.* heat) changes from elongation of rubber is not easy, but may be done in a qualitative way. Stretching a piece of rubber, followed by applying to the lips, which are very heat sensitive, shows the elongated rubber to be warm compared with its temperature before stretching. In other words the stretching process is exothermic; ΔH_{ext} is negative. However, releasing the stress causes an instantaneous return to the unstressed state, with a corresponding reduction in the temperature of the specimen. The free energy change, ΔG_{ext} is thus seen to be against the process of elongation. ΔG_{ext} is therefore positive. From the well-known Gibbs equation,

$$\Delta G_{ext} = \Delta H_{ext} - T\Delta S_{ext} \tag{7.18}$$

it follows that the $T\Delta S_{ext}$ term (and hence ΔS_{ext} itself) must be negative and numerically larger than the ΔH_{ext} term. This in turn leads to the conclusion that the entropy of the final state (*i.e.* the fully extended rubber) must have decreased relative to the initial unextended state. The elongated rubber is thus more ordered than in the unstretched state and the driving force for the rapid return to this less ordered state on removal of the stress is almost entirely the need to maximise the entropy of the system.

This thermodynamic behaviour is consistent with stress-induced crystallisation of the rubber molecules on extension. Such crystallisation would account for the decrease in entropy, as the disorder of the randomly coiled molecules gave way to well-ordered crystalline regions within the specimen. X-Ray diffraction has confirmed that crystallisation does indeed take place, and that the crystallites formed have one axis in the direction of elongation of the rubber. Stressed natural rubbers do not crystallise completely, but instead consist of these crystallites embedded in a matrix of essentially amorphous rubber. Typical dimensions of crystallites in stressed rubber are of the order of 10 to 100 nm, and since the molecules of such materials are typically some 2000 nm in length, they must pass through several alternate crystalline and amorphous regions.

One effect of this strain-induced crystallisation is that there is a characteristic upswing in the plot of stress against strain for natural rubbers, as illustrated in Figure 7.11.

Density is also found to increase in this region, thus providing additional evidence of crystallisation. Certain synthetic elastomers do not undergo this strain-induced crystallisation. Styrene–butadiene, for example, is a random copolymer and hence lacks the molecular regularity necessary to form crystallites on extension. For this material, the stress–strain curve has a different appearance, as seen in Figure 7.12.

When rubbers eventually fracture, they do so by tearing. Fracture surface energies, using the Griffith equation, have been found to be of the order of $10^3 \, \mathrm{J\,m^{-2}}$, whereas the true surface energies are only 0.1–$1.0 \, \mathrm{J\,m^{-2}}$. Hence, more energy is involved in fracture than is required to form new surfaces, and, as with other polymers, this extra energy is assumed to be used up in viscoelastic and flow processes that occur between the molecules immediately before the rubber breaks.

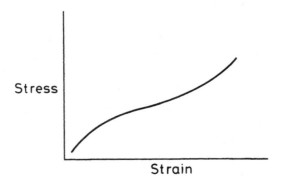

Figure 7.11 Stress relaxation curve for vulcanized natural rubber showing characteristic upswing at higher stresses.

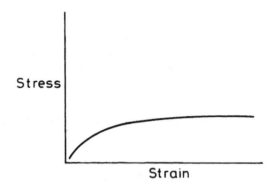

Figure 7.12   Stress–strain curve for styrene–butadiene rubber.

7.12   Reinforced Polymers

The mechanical properties of pure polymeric materials are often inadequate for particular applications, and to overcome this problem these materials may be reinforced in some way. The most common method is to include a substantial amount of a rigid filler or fillers, generally as finely divided powder, or as rods or fibres. For certain materials, elastomeric particles may be used, and these have the effect of reducing brittleness.

In either case the resulting material is a composite, with the polymer as the continuous phase or matrix, binding together the pieces of the discontinuous filler phase. The presence of filler can have a profound effect on the properties of the polymer composite, as illustrated in Table 7.1. From this Table, it can be seen that the nature of the filler is important, with different effects being obtained with different fillers.

Fabricating polymers into composites leads to improvements in their physical properties. In principle, the properties that result from forming composites are a sort of weighted mean of those of the components of the composite. This suggests that to reinforce polymers, mechanically strong substances should be used as fillers, typically inorganic materials such as powdered quartz, carbon black or glass fibres. The composite material that results from blending filler into polymer has properties that lie somewhere between those of the matrix polymer and that of the filler and this is known as the "Rule of Mixtures".

We can see how this applies by considering the elastic modulus. If we have a simple composite consisting of polymer and filler only,

Table 7.1 The effect of particulate fillers on tensile properties of poly(ethylene) (filler concentration 25 parts/100 of polymer by weight).[a]

Filler	Yield stress/MPa	Elongation at yield/%
None	22.2	21
China clay	25.5	26
Carbon black	27.1	19
Barium sulfate	20.9	24

[a]Data from *Composite Materials*, ed. L. Holliday, Elsevier Publishing Company, Amsterdam, 1966.

where V_f is the volume fraction of the filler, then the simplest expression of the elastic modulus of the composite, E_c, is:

$$E_c = (1 - V_f)E_{c}, + V_f E_f \qquad (7.19)$$

where E_p and E_f are the elastic moduli of the polymer and the filler respectively.

This equation turns out to be a not very accurate representation of the elastic modulus even of a simple binary composite. This is because it ignores co-operative effects between the filler and the polymer and between the filler components, and also neglects any effects due to the morphology of the filler. However, it does provide an indication of the way in which the properties of a filled polymer composite arise from the properties of the individual components, and also their respective amounts as measured by the volume fraction.

One of the most important reinforcing fillers is glass fibre, which is used particularly with unsaturated polyester resins. This combination creates an important group of composites which find extensive technical application, for example as boat hulls or as baths and shower units. In these materials, the glass fibres are generally present in the form of chopped strands, the strands being woven together into a mat. The complete composite is made by working the unsaturated polyester, together with its styrene crosslinking agent, into this mat and allowing the styrene to polymerise and also react with the double bonds of the polyester to form the solid matrix.

Being thermosets, styrene crosslinked polyesters tend to fail mechanically by brittle fracture under most test conditions. Typical tensile strengths for the unfilled polymer lie between 40 and 90 MPa and the inclusion of glass fibre leads to improvements in these values. The glass fibres themselves have very high tensile strengths, which may exceed 3000 MPa. This tends to drop during processing, since prior to incorporation into the polymer matrix they are coated with binder, then during incorporation they experience some mechanical damage as the liquid pre-polymer is worked in. Nonetheless, it is estimated that once the glass-fibre reinforced composite is fully formed, the filler is capable of carrying some twenty times the stress of the matrix and this is the origin of the reinforcing effect of these fibres.

Elastomers, of which vulcanized natural rubber is the most important example, also undergo dramatic changes in mechanical properties when filled with particulate solids. In part, knowledge of this particular type of system has been developed empirically as the technology of car-tyre manufacture has advanced.

Carbon blacks are the most widely used fillers for elastomers, especially vulcanised natural rubber. They cause an improvement in stiffness, they increase the tensile strength, and they can also enhance the wear resistance. Other particulate fillers of an inorganic nature, such as metal oxides, carbonates, and silicates, generally do not prove to be nearly so effective as carbon black. This filler, which comes in various grades, is prepared by heat treatment of some sort of organic material, and comes in very small particle sizes, *i.e.* from 15 to 100 nm. These particles retain some chemical reactivity, and function in part by chemical reaction with the rubber molecules. They thus contribute to the crosslinking of the final material.

The addition of filler to synthetic elastomers may lead to significant increases in tensile strength. For example, butadiene or neoprene rubbers may experience a ten-fold increase in tensile strength in the filled state, typically from about 3.5 to 35 MPa.

However, not all properties are improved by filler. One notable feature of the mechanical behaviour of filled elastomers is the phenomenon of stress-softening. This manifests itself as a loss of stiffness when the composite material is stretched and then unloaded. In a regime of repeated loading and unloading, it is found that part of the second stress–strain curve falls below the original curve (see Figure 7.13). This is the direct opposite of what happens to metals, and the underlying reasons for it are not yet fully understood.

Thermoplastic polymers, such as poly(styrene) may be filled with soft elastomeric particles in order to improve their impact resistance. The elastomer of choice is usually butadiene–styrene, and the presence of common chemical groups in the matrix and the filler leads to improved adhesion between them. In a typical filled system, the presence of elastomeric particles at a level of 50% by volume improves the impact strength of a brittle glassy polymer by a factor of between 5 and 10.

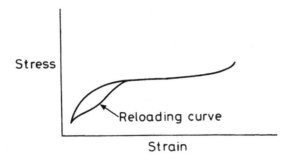

Figure 7.13 Typical stress–strain curve for a filled elastomer.

One reason advanced to account for this improvement is as follows. Impact strength is essentially a function of how readily cracks can be propagated within the matrix. When elastomeric particles are present as filler, they stretch as the crack passes by, thus dissipating much of the energy necessary to develop the new surfaces of the growing crack. The effect of a large number of such particles is to dissipate a large amount of energy. This, in turn, makes crack propagation more difficult, leading to the increased impact strength.

It should be noted that not every kind of strength of a polymer is improved by the addition of elastomeric particles. While impact strength goes up with increasing amounts of filler, tensile strength goes down. This unfortunately is generally true for composites; measures which lead to an increase in impact strength usually do so at the expense of tensile strength.

7.13 Practical Measurements of Mechanical Properties

In order to evaluate the mechanical properties of polymers, specimens are tested in one of two modes, either by applying forces in a single direction, or by applying oscillating forces. The first mode of tests includes measurement of such parameters as tensile strength, compressive strength, or creep behaviour. The second group, the dynamic tests, are used to characterise properties arising from the viscoelastic nature of the polymers, such as storage and loss modulus, and also, from the position of maximum damping as the sample is heated, the glass transition temperatures.

The first group of tests is carried out on specimens generally fabricated into a dumb-bell shape, with forces applied uniaxially. The usual apparatus consists of a machine with a pair of jaws, which during the test are moved relative to each other, either together or apart, in a controlled manner. A chart recorder is employed to give a permanent record of the results obtained, so that the force at fracture can be determined. Whether this kind of set up measures tensile, compressive, or flexural strength depends on how the sample is oriented between the jaws, and on the direction that the jaws are set to travel relative to one another.

All of these tests, by their nature, need to be repeated several times with different specimens of any polymer sample, in order to ensure that there is enough information for statistical analysis. Although the physicist Lord Rutherford said that if your results need statistics, you ought to have done a better experiment, his dictum cannot be extended to tests on mechanical properties of polymers. Such tests can be influenced by a variety of parameters, including surface flaws, internal voids, badly dispersed filler, or other serious imperfections which may lead to marked non-uniformity of specimens. Thus, it is vital to carry out several determinations of strength of a polymeric material, in order that the mean and the scatter of values may be determined. Following this, it becomes possible to determine whether measured differences which may follow changes in sample preparation and/or composition lead to real (*i.e.* statistically significant) differences in the properties of the final material.

Dynamic mechanical tests, which are the other major group of testing techniques, tend not to be subject to such wide variation in results as non-dynamic ones, and hence statistics tend not to be used to the same extent.

This second group of tests is designed to measure the mechanical response of a substance to applied vibrational loads or strains. Both temperature and frequency can be varied, and thus contribute to the information that these tests can provide. There are a number of such tests, of which the major ones are probably the torsion pendulum and dynamic mechanical thermal analysis (DMTA). The underlying principles of these dynamic tests have been covered earlier. Such tests are used as relatively rapid methods of characterisation and evaluation of viscoelastic polymers, including the measurement of T_g, the study of the curing characteristics of thermosets, and the study of polymer blends and their compatibility. They can be used in essentially non-destructive modes and, unlike the majority of measurements made in non-dynamic tests, they yield data on continuous properties of polymeric materials, rather than discontinuous ones, as are any of the types of strength which are measured routinely.

Overall, in order to characterise a polymeric material completely, data from both kinds of test are needed.

Further Reading

Most suggestions for further reading are given in full in the Bibliography at the end of this book. However, in view of the specialised nature of the topics discussed in this chapter it is worth citing a list of further reading which concentrates on mechanical properties of polymers. They are:

L. E. Neilson, *Mechanical Properties of Polymers*, Literary Licensing LL, USA, 2012.

I. M. Ward and J. Sweeney, *Mechanical Properties of Solid Polymers*, John Wiley and Sons, Chichester, 3rd edn, 2013.

8 Polymer Degradation

8.1 Introduction

Organic polymers are not indefinitely stable under all conditions. In particular in fires or on exposure to outdoor conditions, especially sunlight, they will tend to undergo reactions that lead to a loss of their desirable properties. This is known as degradation and often involves both reaction with oxygen and reduction in molar mass. Polymer degradation both in fires and on weathering, together with techniques for minimising the damage caused by these conditions, are the subjects of this chapter.

8.2 Behaviour of Polymers in Fires

Conditions in fires are characterised by intense heat. At the beginning of a fire there is usually a plentiful supply of oxygen, but this can rapidly diminish as the fire proceeds, depending on how well ventilated the fire is. For example in one simulation of a real fire a variety of polymeric items including furniture, bedding materials, settees, and arm-chairs were burnt. These were chosen to represent as closely as possible a fire that had occurred in a department store and which had resulted in 10 deaths. The items were stacked, as they had been in the store, and ignited. One minute after ignition, a temperature of 1000 °C was recorded in the stack; the concentration of oxygen fell rapidly in the vicinity of the stack, reaching 6.5% after 2 minutes and a final level of 3% during the period of sustained combustion.

The Chemistry of Polymers, 5th Edition
By John W. Nicholson
© John W. Nicholson 2017
Published by the Royal Society of Chemistry, www.rsc.org

This example illustrates the range of oxygen concentrations and the kind of temperature that can occur in a real fire, but this is only an illustration. There are no standard conditions for real fires, since each fire is a unique physico-chemical event.

With the increasing use of polymers in both the home and the workplace, there seems to have been a change in the nature of fires. Fire brigades now report fires that are shorter and more intense than previously; there is also much more smoke and significantly greater amounts of toxic gases. All of these arise from the nature of the polymers being used in everyday life.

Combustion of polymers occurs in two broad stages. Firstly there is initiation in which a nearby source of heat causes the temperature to rise to such a point that chemical bonds begin to break. This generates low molar mass products that migrate through the polymer and out into the gas phase. There they undergo oxidation, *i.e.* they burn. This gas-phase oxidation is the second stage of the process. Heat produced from this stage is fed back to the bulk polymer causing further breakdown of the molecules to yield more volatiles as fuel for the combustion stage. For many polymers this whole cycle becomes self-sustaining and continues until no further volatile products can be generated from the polymer.

Thermosets and thermoplastics behave differently from each other in fires. Thermosets do not melt when heated but may well undergo further crosslinking. The presence of such additional crosslinks hinders movement of any volatile degradation products through the polymer matrix. Hence the combustion zone tends to be starved of fuel and for this reason thermosets tend to be relatively non-flammable.

Thermoplastics, on the other hand, melt when heated and this assists the volatiles to move through the material. Thus fuelling of the combustion zone is readily achieved and, for this reason, many thermoplastics are highly combustible unless treated appropriately.

Polymers do not burn with complete efficiency, even those composed only of carbon, hydrogen, and possibly oxygen. Instead they give off significant amounts of smoke and leave behind a solid residue of char. The quantity of both smoke and char produced may vary widely, depending on both the fire conditions and the material involved. Smoke is a complex mixture of liquid and solid hydrocarbons, oxygenated organic species, carbon particles and, in some cases, water droplets. Char is a brittle solid of variable composition, though generally with an extremely high carbon content.

8.2.1 The Behaviour of Individual Polymers

A number of thermoplastics undergo depolymerisation on heating. These include poly(styrene), poly(methyl methacrylate), and poly(oxymethylene). Such depolymerisation will occur regardless of the prevailing oxygen concentration and under well aerated conditions will provide a ready source of fuel for sustained combustion.

PVC behaves differently in fires. It readily loses HCl from along the back bone in a process that is autocatalytic. The product left behind has a highly conjugated structure that gives rise to a deep orangy-red or black colour and makes the product very brittle. The hydrogen chloride given off is not, of course, flammable so that despite extensive degradation PVC never undergoes self-sustaining combustion.

Dehydrochlorination of PVC will occur at the kind of temperatures used in the commercial processing, making it the least thermally stable of all the major tonnage polymers. Because of this PVC always contains a stabiliser, the precise nature of which will depend on the proposed end-use of the particular batch of PVC. The stabilisers used to protect against fire damage are different from those used to protect against degradation in processing. Still others are used when protection against photolytically induced dehydrochlorination is needed. The range of stabilisers may also be influenced by other considerations; for example, PVC film for use in food packaging has to be stabilised with an additive that will not migrate into the food or taint at the point of contact.

8.3 Assessment of Combustion Behaviour

An important technique for evaluating the combustion behaviour of polymers is determination of the Limiting Oxygen Index (LOI) of the material. The use of the LOI is a semiquantitative method that was developed originally by Fenimore and Martin in 1966 for the general evaluation of organic substances, not only polymers. In the technique, which uses polymer specimens of a standard size, various oxygen concentrations in an oxygen–nitrogen mixture are used and the minimum amount that will just allow flaming combustion to continue for at least 3 minutes is determined. This value, expressed as a percentage, is known as the Limiting Oxygen Index. Typical results are shown in Table 8.1.

Values for LOI are found to be reasonably reproducible and they give some indication of flammability. However, their use in this

Table 8.1 Limiting Oxygen Index of some polymers.

Polymer	LOI/%
Poly(methyl methacrylate)	17.3
Poly(propylene)	17.4
Poly(ethylene)	17.4
Poly(styrene)	18.1
Poly(vinyl alcohol)	22.5
Poly(vinyl chloride)	45.0
Poly(vinylidene chloride)	60.0
PTFE	95.0

respect is limited, since the conditions in real fires differ in important respects from the conditions of the LOI test. For example, real fires often generate a considerable velocity of air, which sweeps away degradation products that may otherwise suppress combustion. Secondly, the temperature of the air in real fires may rapidly become very high. Hot, well-oxygenated air is more likely to initiate substantial damage to organic polymers than the gaseous mixture used in the LOI test. The use of an artificially low temperature reduces the apparent flammability of the measured LOI and hence may over-estimate the safety of a given polymer under fire conditions. Because of these limitations data from the LOI tests need to be supplemented by results from other test methods.

8.4 Improvement of Stability of Polymers in Fires

The major improvement sought in polymers in terms of their fire behaviour is reduction of flammability. For certain applications, however, reduction in smoke evolution is sought but these two aims tend to be mutually incompatible. Reduction in flammability is brought about by making the combustion process less efficient. A penalty for inefficient combustion is increased smoke production. Similarly a reduction in smoke evolution may be achieved by increasing the efficiency of any accidental combustion; that is, by increasing the flammability.

Reduction in flammability is achieved by the incorporation of flame retardants into the polymer. Two possible approaches to this are available; either the use of additives blended into the polymer at processing stage (additive type) or the use of alternative monomers which confer reduced flammability on the final product (reactive type). A number of elements have been found to assist with conferring flame retardancy on polymers, the main ones being bromine, chlorine, nitrogen, and phosphorus.

A large number of materials, generally inorganic in nature, have been used as additive-type flame retardants, of which hydrated alumina and antimony oxide are among the most important.

Flame retardants interfere with one or more of the essential processes of self-sustaining combustion. For example inorganic materials of good thermal conductivity may improve the rate of heat transfer throughout the polymer. This reduces the local temperature at the surface close to the ignition source, which in turn reduces the rate and extent of degradation reactions taking place within the polymer. The use of certain glassy solids, such as metal borates, may lead to the formation of an impervious glassy layer which prevents the passage of volatiles out to the combustion zone. Finally the incorporation of chlorinated or brominated additives may give rise to substantial quantities of either hydrogen halide or halogen gas within the volatiles, thus quenching vapour phase oxidation in the combustion zone.

In creating less flammable grades of polymer use is often made of the so-called antimony–halogen synergism. Synergism refers to the situation in which the joint effect is greater than the sum of the individual effects and in the case of antimony–halogen synergism this means that smaller amounts of each material can be used than would have been expected on the basis of the individual effects of adding Sb_2O_3 or incorporating co-monomers containing bromine or chlorine.

The mechanism of antimony–halogen synergism is believed to be:

(i) Polymer, in combustion conditions, begins to evolve HCl or HBr;
(ii) HCl or HBr reacts with Sb_2O_3 to give SbOX;
(iii) SbOX undergoes self-reaction to yield SbX_3, a gas, plus Sb_2O_3;
(iv) SbX_3 quenches the flame by acting as a radical trap and by preventing access to oxygen. As such, it is more effective than either of the hydrogen halides or the halogen gases as an agent for quenching combustion.

There is good experimental evidence for this mechanism of antimony–halogen synergism, most significantly that flame extinction time is a minimum for polymers containing antimony and halogen in the mole ratio 1 : 3.

8.5 Weathering of Polymers

Weathering is a broad term that is applied to the changes that take place in a polymer on exposure out of doors. The main agents of weathering are sunlight (particularly ultraviolet radiation), temperature, thermal cycling, moisture in various forms, and wind. The main degradation is brought about by ultraviolet light, assisted by contributions from the visible and near-infrared portions of the electromagnetic spectrum. In particular the near-infrared radiation accelerates degradation reactions by raising the temperature.

All of the factors involved in weathering, including both the amount of intensity of sunlight, vary both seasonally and geographically. To understand fully and predict the weathering behaviour of any polymer requires information about exactly how these factors vary and how they then contribute to the overall degradation process.

Solar radiation experienced by polymers at the Earth's surface has two components, direct radiation from the sun and indirect radiation from the sky. At all parts of the Earth's surface, solar irradiance is highest at noon, when the sun is highest in the sky. The actual intensity, though, varies geographically. This is due not only to regional variations in the amount of direct sunlight, but also because the clarity of the atmosphere varies geographically, depending on the amount of water vapour or polluting gases that are present locally.

One part of the world where solar irradiance is very strong is in Florida. High temperatures, direct sunlight, and relative absence of pollution all contribute to this strong irradiance and because of it there are many sites in Florida that are used to test the resistance of polymers to outdoor exposure. Typically banks of racks holding polymer or paint samples inclined at 45° are exposed to the sun for several months or even years. The extent of photodegradation may then be assessed, often fairly subjectively, based on the understanding that polymer samples which survive in the high irradiance of Florida will tend to do well almost anywhere else in the world.

Sunlight alone is not responsible for the degradation of polymers; the process is assisted by the presence of oxygen so that photo-oxidation is able to occur. Oxygen can promote polymer degradation in the presence of solar radiation in a number of ways. Free radicals formed in the polymer by photolysis can react with oxygen to form peroxy radicals which initiate a series of radical chain reactions. These processes occur by reaction with oxygen in the ground state. However, reaction of ground-state oxygen with sensitisers, such as dyes in the excited state, causes the formation of excited-state oxygen. In terms of electronic spin states, this generates the more reactive singlet oxygen from the less reactive triplet oxygen. Singlet oxygen is responsible for the deterioration in sunlight of unsaturated polymeric materials such as natural rubber and synthetic elastomers.

Polymers are rarely used alone in applications involving outdoor exposure but are usually either filled or pigmented. Pigments or fillers may themselves influence the course of photo-oxidative degradation. For example the rutile form of titanium dioxide is the most widely used white pigment for both polymers and paints and it may either assist or inhibit degradation. TiO_2 may assist degradation because it acts as a photosensitiser. This occurs by the following mechanism:

$$TiO_2 \rightarrow \text{(electron–positive hole pair)}$$
$$\text{(electron–positive} + Ti^{4+} \rightarrow Ti^{3+} + \text{positive hole}$$
$$\text{hole pair)}$$
$$O_2 + e^- \rightarrow O_2^-$$
$$O_2^- + \text{positive hole} \rightarrow \text{singlet } O_2 \text{ or } 2O.$$

Hence either the reactive singlet oxygen molecule or oxygen atoms are produced; either of these may initiate radical chain processes that lead to degradation.

Alternatively because of the degree to which it both reflects white light and absorbs in the ultraviolet, it may inhibit photodegradation. Which of these two possibilities predominates depends on a variety of factors including the concentration of the TiO_2 pigment used, the nature of any surface treatments, and the extent to which the polymer itself is able to resist photodegradation.

The testing of polymers for resistance to weathering is difficult. Apart from exposing samples in places that naturally experience high solar irradiance, such as Florida, various artificial weathering devices are available. It must be stressed that these are *artificial* weathering devices. They cannot accelerate weathering, because the factors which cause weathering are complex and co-operative. Hence they cannot be readily accelerated.

There is great difficulty in correlating results obtained from artificial weathering tests with those involving outdoor exposure. Part of the problem is that it is not possible to mimic solar radiation completely. Solar radiation comprises a spread of wavelengths, with intensities varying according to wavelength. Artificial weathering using arc lamps cannot completely reproduce both the spectral spread and the variation in intensity, even though the basic requirements of heat, ultraviolet light, and moisture may be met.

Another difficulty with testing is that exposure of specimens at a site such as Florida may give a different pattern of degradation from tests involving exposure at a site having less incident sunlight. Interpretation of the results of any weathering test, whether artificial or natural, requires care and experience. In the end there is no substitute for seeing how the polymer performs when exposed to the conditions under which it will eventually be used.

8.6 Protection of Polymers From Photo-oxidation

A variety of additives are available for incorporation into polymers to act as antioxidants. A selection of these additives is listed in Table 8.2 together with the names of the polymers for which they are most suited.

Whilst not an exhaustive list the compounds given in Table 8.2 do represent the major classes of antioxidant. One feature that is clear from this table is that these antioxidants tend to be effective in many different polymers, showing that they operate by interfering with the agents of photo-oxidation rather than by a specific interaction with the particular polymer.

One general mechanism by which antioxidants act is by reaction with peroxy radicals. In doing so they compete with the polymer, thus reducing the extent to which the degradation mediated by peroxy radicals can occur. In the presence of appropriate hydrogen-donating

Table 8.2 Antioxidants used in polymers.

Compound	Suitable polymers
2,6-Di-*t*-butyl-4-methyl phenol	Polyamides, poly(esters), poly(alkenes), PVC, polyurethanes, and rubbers
2,5-Di-*t*-phenyl hydroquinone	Poly(esters) and rubbers
N,N'-Diphenylphenylenediamine	Poly(alkenes) and rubbers
Didodecyl 3,3'-thiopropionate	Cellulosic polymers, ethylene–vinyl acetate copolymers, poly(alkenes), PVC, rubbers, and poly(styrene)
Tris(nonylphenyl) phosphite	Cellulosic polymers, poly(esters), poly(alkenes), polyurethanes, PVC, rubbers, and poly(styrene)

antioxidants peroxy radicals undergo one of two propagation steps (Reaction (8.1) or (8.2)).

$$ROO\cdot \quad + \quad AH \quad \longrightarrow \quad ROOH \quad + \quad A\cdot \qquad (8.1)$$

$$ROO\cdot \quad + \quad RH \quad \longrightarrow \quad ROOH \quad + \quad R\cdot \qquad (8.2)$$

The rate constant of Reaction (8.1) is much greater than the rate constant of Reaction (8.2), which means that antioxidants of this type can be used in very low concentrations with good effect. A typical thermoplastic would contain only 0.01–0.5% by mass of such an antioxidant. Typical compounds which work by this mechanism include substituted phenols and secondary aromatic amines.

An alternative mechanism by which additives may protect polymers from photo-oxidation is radical trapping. Additives which operate by this mechanism are strictly light stabilizers rather than antioxidants. The most common materials in this class are the hindered amines, which are the usual additives for the protection of poly(ethylene) and poly(propylene). The action of these stabilisers is outlined in Reactions (8.3)–(8.5).

$$>\!\!NH \quad \xrightarrow{\text{radicals, } O_2} \quad >\!\!NO\cdot \qquad (8.3)$$

$$>\!\!NO\cdot \quad + \quad R\cdot \quad \longrightarrow \quad >\!\!NOR \qquad (8.4)$$

$$>\!\!NOR \quad + \quad ROO\cdot \quad \longrightarrow \quad >\!\!NO\cdot \quad + \quad ROOR \qquad (8.5)$$

These additives are thus able to trap both alkyl and peroxy radicals. In this way they interfere with the propagating steps of the degradation process. Since overall the nitroxyl radicals are not consumed in this mechanism these additives are effective at low concentrations in the polymer.

The final possible mode of action for an antioxidant is as a peroxide decomposer. In the sequences that lead to photodegradation of a polymer the ready fragmentation of the hydroperoxide groups to free radicals is the important step. If this step is interfered with because the peroxide has undergone an alternative decomposition this major source of initiation is removed. The additives which act by decomposing hydroperoxide groups include compounds containing either divalent sulfur or trivalent phosphorus. The mechanism involves expansion of the valence state of the key element to accommodate oxygen, as illustrated in Reactions (8.6) and (8.7).

$$R_2S \ + \ ROOH \ \longrightarrow \ R_2S{=}O \tag{8.6}$$

$$R_3P \ + \ ROOH \ \longrightarrow \ R_3P{=}O \tag{8.7}$$

Depending on the nature of the sulfur or phosphorus compound used, the product $R_2S{=}O$ or $R_3P{=}O$ may undergo a number of further reactions with ROOH groups, all of which convert the hydroperoxide group into an alcohol. These compounds tend to be only weakly effective so are generally used in conjunction with synergistic promoters. Suitable mixtures are used to stabilise a variety of polymers including poly(alkenes), ABS, and poly(styrene).

Compounds for use as antioxidants have to fulfil a number of requirements in addition to their effectiveness in stabilising the polymer. For example, they must have low toxicity and be inexpensive. They must also be fully compatible with the polymer of choice since otherwise they tend to migrate to the surface where they impart an unacceptable appearance. Any such migration also depletes the concentration in the bulk which leads to a loss of protection for the polymer.

8.7 Biological Degradation of Polymers

The major commercial polymers are the polyolefins, poly(ethylene) and poly(propylene), and these are extremely resistant to biological degradation. Some degree of biodegradability can be conferred by copolymerising the principal monomer with a more polar monomer, typically an ester. Alternatively, blending the polyolefin with a highly biodegradable polymer, such as starch, can be employed. Such materials do not fully degrade since the polyolefin continues to resist such degradation. However, the product of degradation tends to be a finely divided powder, rather than large pieces of polymeric debris, and this may be less obtrusive in the environment.

Synthetic polyesters are capable of being degraded by microorganisms, polyesters with aliphatic backbones being more susceptible to such degradation than those with aromatic backbones. Polyamides of low molar mass, too, may be degraded by microorganisms, but commercial polyamides, the nylons, tend to be of too great a molar mass and hence are almost completely resistant to biological attack.

On the other hand there are polymers that are capable of being degraded, generally in a controlled way, to yield low molar mass molecules that can be safely metabolised by living organisms. Polymers of this type need to have relatively polar substituents, preferably ester and hydroxyl groups. One such polymer that has achieved some technical success, for example as a material for biodegradable packaging, is poly(hydroxybutyrate) (PHB).

Biodegradation is an enzyme-catalysed process, and typically occurs in two stages. In the first of them, the enzyme binds to the polymer, after which, in the second stage, hydrolytic cleavage occurs. Generally the more hydrophilic the polymer, the more susceptible it is to hydrolysis. However, other factors also affect the susceptibility towards hydrolysis, such as molar mass and crystallinity. Lower molar mass polymers degrade quicker than their higher molar mass analogues, and amorphous regions of semicrystalline polymers degrade more readily than the crystalline regions.

In nature, there are several sources of enzymes that are capable of catalysing the hydrolysis of PHB. The polymer itself is produced by bacteria and occurs in cells as discrete inclusion bodies. These bodies contain the necessary enzymes for degrading the polymer, preventing its build-up in the cell. As well as this, there are numerous bacteria and fungi, many of which are found in the soil, that are capable of secreting the necessary enzymes outside their cell walls, and thus of initiating degradation of PHB.

Other types of biodegradable polymers include the various polymers used in surgery as degradable sutures, *e.g.* poly(lactic acid) in its various structural isomers, and poly(glycollic acid). These degrade readily within the body, generating low molar mass and monomeric fragments, which are rapidly transported away from the site of the body in which they are placed, and either metabolised or exceted. As with PHB in the soil, degradation is enzymic, but the enzymic route may not be the only degradation mechanism involved. The reactive hydroxy radical HO·, which may be formed as a byproduct of the activity of phagocytosing macrophages, has also been shown experimentally to cause the hydrolytic degradation of the DL-isomer of poly(lactic acid).

Biodegradable polymers are likely to be increasingly important materials in the future, finding use in applications as diverse as medicine, agriculture, and pharmacy. For applications such as packaging, they remain expensive. However, with changing public attitudes towards environmental pollution, it is likely that objections based purely on cost will diminish, and that such applications will also grow in the years ahead.

8.8   Exploitation of Polymer Degradation

The earlier sections of this chapter have assumed that polymer degradation is undesirable, which is true to a large extent. However, there are two related areas of technology, lithography and microlithography, which make positive use of polymer degradation and for which polymers are designed to be as readily degraded by light as possible.

Polymers that are sensitive to light have been used in lithography for some years. Lithographic plates which are based on polymer films are used for printing newspapers and magazines. These polymer films are exposed and developed to form the image of the text to be printed. Ink is then applied to this image-bearing film and because of differences in the affinity for the ink caused by exposure to light the ink is taken up by the plate selectively. This selective ink uptake is responsible for forming the image as the ink is applied to the paper.

Related to this technology is the technique of microlithography. This latter technique is used for making the images on microchips that form the components of integrated circuits. Photodegradation of polymers is also a crucial part of this process, but greater precision is needed than for conventional lithography, since modern random access memory chips have features in the 1 μm size range (compare this with the thickness of a typical coat of paint at 25 μm or the diameter of a human hair at 70 μm).

To produce the images necessary for use in an integrated circuit by the process of microlithography a silicon wafer is first coated with a light-sensitive polymer. A mask is placed over this bearing the image that is required to be reproduced on the silicon wafer. The whole arrangement is then exposed to light of the appropriate wavelength for a suitable length of time. This exposure causes some kind of chemical change in those areas of polymer not covered by the mask (*i.e.* either degradation or crosslinking).

The next stage of the process is called development. In this stage the image of the mask is developed in the polymer by treating the polymer with solvent. The precise mode of development depends on the type of photopolymer used. So-called positive photoresists have undergone degradation on exposure to light and so dissolve at those points not covered by the mask. This leaves polymer behind in the shape of the mask. Subsequent etching of the silicon wafer leads to the formation of an image of the mask that stands proud of the surface. Removal of the polymeric photoresist leaves this positive image behind on the silicon wafer.

Alternatively the photopolymer can crosslink under the influence of light, thereby becoming less soluble. These polymers are known as negative photoresists and they will dissolve away in appropriate solvents only from those regions not originally covered by the mask. Treatment of the silicon with etching solution will then be able to remove material from the surface at those points which result in an image of the mask that lies below the level of the main surface. Completion of the process by removal of the polymer results in the formation of a negative image of the mask on the silicon wafer.

An outline of these two possible processes of microlithography is shown in Figure 8.1.

In order to be successful as a photoresist the polymer must have three characteristics:

1. It must be sensitive to the desired wavelength of light, which in conventional manufacturing practice is 436 nm, though improvements follow if shorter wavelengths are used.
2. In addition to this high sensitivity the polymer must be capable of giving high-resolution images.
3. The polymer must resist the etching solutions used to remove the unwanted surface layers of the silicon wafer as the image is transferred to the microchip.

A number of polymers are capable of fulfilling these demanding requirements. Typically negative photoresists are based on cyclised poly(1,4-isoprene). These polymers are prepared by dissolving poly(1,4-isoprene) in an appropriate solvent and subjecting it to thermal degradation. This is followed by treatment with acid to produce the cyclised material (see Reaction (8.8)).

$$(8.8)$$

These polymers need to be made photosensitive for use as photoresists and this is achieved by the incorporation of bisazide sensitisers. On exposure to light the photochemical reaction induced by the bisazide results in rapid crosslinking of the polymer rendering it insoluble in the developing solvent. More recently, poly(vinylcinnamate) has also been used as a negative photoresist. The cinnamate functional group undergoes a $[2+2]$ cycloaddition when exposed to light, which crosslinks the molecules, rendering them insoluble.

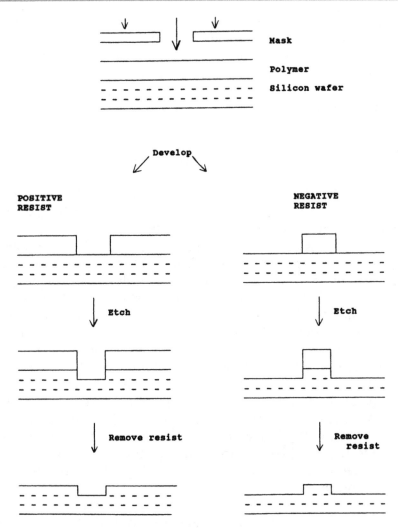

Figure 8.1 Principle of microlithography.

Positive photoresists, by contrast, are based on water-soluble no-volak resins with naphthalene diazoquinone sulfonate (NDS) as the photosensitiser. On photolysis the NDS causes a rearrangement in the polymer to yield nitrogen gas plus an indene carboxylic acid. This latter functional group considerably increases the solubility of the polymer, hence solubilising those areas of the polymer that had been exposed to light.

These examples are relatively well established but new polymers for use as photoresists are continually appearing. A major driving force for these developments is the need to use shorter wavelengths (below

193 nm) in order to obtain finer features in the resulting device. Novolaks, being aromatic, are opaque at these wavelengths, so are not suitable. Hence, new polymers are needed which are capable of providing the necessary properties. The research literature contains examples of these newer systems and this remains are active area of current research.

8.9   Polymer Light Emitting Diodes

A light emitting diode (LED) is a device which emits light in response to an electric current. These devices find use in television screens, computer monitors and mobile phones. Polymer LEDs emit less light than inorganic LEDs, but they are mechanically flexible, and hold the promise of wide application as their technology is developed.

Electroluminescence from films of poly(n-vinylcarbozole) was first demonstrated in 1975 by Roger Partridge at the UK's National Physical Laboratory. The polymer was in the form of a thin layer approximately 2 μm thick between two plates which acted as electrodes.

This arrangement is the same as in modern LEDs, because at least two layers are needed for these devices to function. One of the layers transports the electrons and the other the holes. Electrons and holes combine at the interface between the layers, and this leads to light emission. Modern devices employ very thin films, typically of the order of 100 nm thick. Such polymer films are stable and capable of being formed into high quality films. The films themselves are typically formed from solutions of polymer, using spin coating to ensure an even and thin film.

The essential requirement of the polymer component of LEDs is that they are electrically conductive. As with those polymers described in Section 10.4, this means that they have to contain delocalised Π electrons that are conjugated over at least part of the molecule. However, conductivity is not necessarily as high as for true conducting polymers, and they may fall into the category of semiconductors.

The typical architecture of a polymer LED is shown in Figure 8.2. In use, a voltage is applied across the polymer LED, causing a current of electrons to flow from the cathode to the anode. This requires electrons to be withdrawn from the lightest occupied molecular orbital of the anode polymer (the emissive layer) and transported to the lowest unoccupied molecular orbital of the cathode polymer (the conductive layer). This is effectively the combination of electrons with holes, the

	Cathode
− − − − − −	Emissive layer (electron-rich)
+ + + + + +	Conductive layer (rich in holes)
	Anode

Figure 8.2 Architecture of polymer LED device.

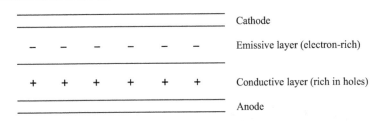

Figure 8.3 Structure of poly(*p*-phenylene vinylene).

result of which is to produce an exciton, the decay of which results in the emission of a photon of light. This light has characteristic energy, and hence colour.

Polymer LEDs typically employ indium tin oxide, which is transparent to visible light, as the anode, and metals such as calcium or barium as the cathode. These cathode metals are reactive, so are capped with a layer of aluminium to preserve them in their active state. Polymers are typically derivatives of poly(*p*-phenylene vinylene) (see Figure 8.3) or polyfluorene, often with side chains. These side chains alter the energy of the light emitted, hence control colour, and may also stabilise the polymer. In addition, side chains may assist dissolution in appropriate solvents, and this improves the ease of fabrication of the polymer LED device.

This is a field in which developments are occurring rapidly. In the near future we are likely to see much thinner devices with considerable flexibility, as well as improved efficiency, stability and light output. These will find application in a variety of display devices, including 3D television screens.

9 Dendrimers

9.1 Introduction

The study of dendrimers is a relatively new branch of macromolecular chemistry and dates from the mid 1980s. Dendrimers are macromolecules that have an extensively branched, spherical structure based on a central core and numerous branches. The outer surface typically includes specific functional groups. The word dendrimer comes from the Greek word *dendra*, which means tree. These substances have also been called *arborols* and *cascade molecules*, but the term *dendrimer* is preferred and is the one used throughout this chapter.

Dendrimers are macromolecules that consist of numerous branches growing out of a central point or core. They are highly monodisperse (typical polydispersities are of the order of 1.02–1.04 but may be as low as 1.001), and they have a globular morphology resulting from the fact that the branches radiating out from the centre of the molecule are of almost identical length. Individual dendrimer molecules comprise three distinct zones, namely the core, the branches and the end-groups.

Dendrimers have distinctive properties, such as the ability to entrap small molecules in their core region and very low intrinsic viscosities in solution. Such properties require molecules to have achieved a particular size, and not all molecules with branches radiating from a core are large enough to develop the characteristic properties of true dendrimers. Branched molecules below this critical size are called *dendrons* and are the equivalent in dendrimer chemistry of oligomers in polymer chemistry.

The Chemistry of Polymers, 5th Edition
By John W. Nicholson
© John W. Nicholson 2017
Published by the Royal Society of Chemistry, www.rsc.org

The reason for the low intrinsic viscosities in solution is that dendrimers exist as tightly packed balls. This is by contrast with linear polymers, which tend to form flexible coils. The effect of this difference is that, whereas polymer solutions tend to be of high viscosity, dendrimer solutions are of very low viscosity. In fact, as dendrimers are prepared, their intrinsic viscosity increases as far as the addition of the fourth monomer unit to growing branches (the so-called fourth generation), but this is the maximum value that the viscosity reaches, and as the side chains grow beyond that, the viscosity decreases.

The presence of a large number of chain-ends in the fully synthesized dendrimer molecules makes them highly soluble and also readily miscible, for example with other dendrimer solutions. The solubility is controlled by the nature of the end-groups, so that dendrimers with hydrophilic groups, such as hydroxyl or carboxylic acid, at the ends of the branches are soluble in polar solvents, whereas dendrimers with hydrophobic end-groups are soluble in non-polar solvents. The density of the end-groups at the surface of the dendrimer molecule means that they have proportionately more influence on the solubility than in linear polymers. Hence a dendritic polyester has been shown to be more soluble in tetrahydrofuran than an equivalent linear polyester.

Dendrimers can be created by one of the two synthetic approaches, which are shown diagrammatically in Scheme 9.1. They are:

(i) Growth from the core, the so-called divergent approach. In this, the future dendrimer core is created first and the branches grown out from it by the use of appropriate techniques of synthesis.

(ii) Growth from the end-groups inwards, the so-called convergent approach. Here, the future periphery of the molecule is prepared first, after which the synthesis proceeds inwards in a stepwise manner until the core is completed.

The choice of synthetic approach is based on a number of factors, such as the desired features in the target dendrimer molecule, the chemistry available for the development of the branching zone and the specific chemical moieties available as building blocks for the dendritic framework. The convergent approach is often better at providing control of the resulting dendritic structure. This is partly due to the relative ease of purification of the intermediate chemical species and partly because the functional groups being used for the inward growth are different from the end-groups, hence there is

Divergent synthesis

Synthesis from the core (X) to the periphery

Convergent synthesis

Synthesis achieved by condensation of molecular fragments to the focal point, Y.

Scheme 9.1

greater uniformity in the resulting structure. It is harder to maintain chemical purity or structural consistency in the divergent approach, because the number of reactions that must be completed grows exponentially with each step, and by the end needs large excesses of reagent to complete the process.

Dendrimers are generally considered to be relatively homogeneous substances. However, that may depend on the route by which they were made. Divergent synthesis often results in a material that is impure or structurally imperfect. If we consider the divergent approach, the reason for this becomes apparent. To obtain a sixth generation dendrimer, which is 99% pure, from a fifth generation containing 96 sites requires the coupling reaction to proceed to 99.99% conversion. This is, at best, a demanding requirement and suggests that few true dendrimers are of particularly high purity.

The convergent approach overcomes these problems to an extent, but still gives more or less imperfect structures depending on the purity of the reagents and the efficiency of the reactions used.

Most dendrimers that have been prepared in the years since 1985 have been based on covalent bonds. However, some have employed secondary bonding, such as hydrogen bonding or co-ordination chemistry. Dendrimers of this type can be considered as part of the growing field of supramolecular chemistry, and can be formed by the kind of self-assembly processes that are typically used in this field.

9.2 Organic Dendrimers

Organic dendrimers were the first to be prepared. The idea of preparing them grew from developments in host–guest and supramolecular chemistry, in particular the efforts to prepare large substrate-selective ligands that were called, graphically, "tentacle" or even "octopus" compounds. As their name suggests, these molecules had long branches radiating from a central hub, though there were only a few of them compared with dendrimers.

Nonetheless, it was a fairly short step from "octopus" compounds to dendrimers, and the step was taken by Vogtle in the late 1970s when he attempted to use a cascade reaction to prepare a molecule of the dendrimer type that would now be considered a dendron rather than a fully developed dendrimer. It began with the addition of acrylonitrile to an amine, followed by reduction of the nitrile to amine. This was followed by a further reaction with acrylonitrile, and the process was repeated several times to yield highly branched macromolecules. There were initially problems with the reduction step but these were overcome, and the preparation of these poly(propylene imine) dendrimers was later commercialized.

The sequential growth and branching involved in the preparation of dendrimers had been considered by Flory many years before they were actually prepared. Flory developed a sound understanding of the kind of processes that would occur in the self-polymerization of a molecule of the type AB_2 most of which have been shown to be correct by the relatively recent experimental studies. In particular, the existence of a limit to growth was predicted. This limit has become known as the *starburst limit*, and is the reason for the highly monodisperse nature of fully developed dendrimers.

This phenomenon can be explained as follows: As the number of synthetic steps in a dendrimer preparation grows, so the number of branch ends increases exponentially. For each additional generation of branches, the surface area of the dendrimer molecule increases by a factor of $(generation)^2$, whereas the volume increases by a factor of $(generation)^3$. It follows from this that the surface region of the growing dendrimer molecule becomes increasingly crowded until, in the limit, it is not possible to fit in any more monomer molecules at the ends of the branches. When this happens, growth stops, effectively for steric reasons, and the molecule remains at that size. Since this size is a constant for any particular dendrimer system, it follows that all molecules eventually reach this size and can get no bigger. The absence of all but the smallest variation between the sizes of the molecules generates a system of highly uniform molecular sizes, in other words, an almost monodisperse system.

The starburst point occurs at a well-defined limit for each dendrimer system, and its occurrence is dependent mainly on (a) the functionality of the core, (b) the multiplicity of the branches and (c) the branch length. However, the volume of the core itself, and the length of the monomer branches also have an influence. The end groups may occupy the outer surface of the dendrimer, or the branches may fold inwards, thus distributing the end-groups within the dendrimer. The factors that control this behaviour of the branches are not fully understood, but include the nature of the solvent and the detailed chemistry of the dendrimer branches.

The first true dendrimers were the polyamidoamines (PAMAMs). They are also known as starburst dendrimers, and the term *starburst* is a trademark of the Dow Chemical Company, who have commercialized these materials for a range of applications. These dendrimers use ammonia as the core molecule, and this is reacted with methyl acrylate in the presence of methanol, after which ethylenediamine is added. This is shown in Scheme 9.2.

By the end of the reaction sequence shown in Scheme 9.2, each branch has a free amine group at the end. Each of these amine groups can react with two molecules of methyl acrylate and two molecules of ethylenediamine. The two-step process to generate further branches that are also amine-terminated is considered a generation, so that in this process, each generation results in amine end-groups and each half-generation in carboxylate ester end-groups. An alternative representation of the dendrimer is shown in Figure 9.1.

The properties of the various PAMAM dendrimer precursors are shown in Table 9.1.

Scheme 9.2

Figure 9.1 Alternative representation of PAMAM dendrimer.

Table 9.1 Properties of PAMAM dendrimers containing an ammonia core.

Generation	Relative molar mass	Number of NH terminal groups
0	359	3
1	1043	6
2	2411	12
5	21 563	96
10	700 091	3072

Lower generation dendrimer precursors have highly asymmetric shapes and have more open structures than the higher generation materials. As dendrimer molecules approach the starburst limit, they become more globular in structure and eventually the end-groups form a close-packed membrane-like covering. For PAMAM dendrimers, the limit to growth is the tenth generation. At this point, when the dendrimer is fully formed and unable to increase in size any further, it contains 6141 monomer units and has a diameter of about 1240 nm.

9.3   Supramolecular Dendrimers

Dendrimers can be constructed from chemical species other than purely organic monomers. For example, they can be built up from metal branching centres such as ruthenium or osmium with multidentate ligands. The resulting molecules are known as metallodendrimers. Such molecules can retain their structure by a variety of mechanisms, including complexation, hydrogen bonding and ionic interactions.

An example of this type of dendrimer is the polynuclear species that can be prepared from 2,3-bis(2-pyridyl)pyrazine as bridging ligands, bipyridines as terminal ligands, and either $Ru(\text{II})$ or $Os(\text{II})$ as the metal centre.

Polynuclear dendrimer complexes of this type can undergo redox reactions at the metal centre and have luminescent properties. They have been proposed as molecular photochemical devices, although no practical examples have yet been produced.

Co-ordination chemistry is not the only possible approach to the successful preparation of supramolecular dendrimers. Other examples include hydrogen-bonded dendrimers, which have been prepared using tetracar-boxylic acid moieties as the linking species, and others have been prepared where the hydrogen bonds were developed using melamine and cyanuric acid at the focal point (*i.e.* the point towards which convergent synthesis is directed).

Ionic interactions have been used to prepare lanthanide-core dendrimers. This has been achieved using a convergent synthesis, in which polyether dendrons with a carboxylic acid group at the focal point were assembled around a lanathanide cation. This involved a metathetical reaction with compounds such as $Er(OAc)_3$, $Tb(OAc)_3$ or $Eu(OAc)_3$ to introduce the appropriate lanthanide ion.

9.4   Shapes and Conformations of Dendrimers

The main question concerning the shape and confirmation of dendrimers is the location of the chain-ends. As we have already seen, these can either be folded in or at the outermost part of the dendrimer globule, the so-called periphery. Which form is adopted depends partly on the degree of rigidity in the branches. True dendrimers, that is to say those with significant extents of branching and with molecular dimensions controlled by the inherent steric growth limit, tend to be fairly rigid and to have a globular structure. Small dendrimers, with relatively few branches and long connectors between the branches, tend to be more readily deformable and capable of adopting flatter conformations that involve a degree of chain folding.

Dendrimers appear to have interiors that are, to all intents and purposes, empty and they, therefore, are able to accommodate guest molecules and also nanoparticles. Early theoretical work suggested that dendrimers develop in concentric shells, and enclose a considerable amount of empty space. More recent theoretical studies have suggested that they may not be as much free space as first thought, and this has been confirmed by X-ray diffraction studies. NMR has shown that there is a reasonable free volume within dendrimers though there is some experimental evidence that the amount of free volume varies with the thermodynamic quality of the solvent. This, in turn, suggests that the amount of free volume is tunable by choosing appropriate solvents or solvent blends.

The cavities in dendrimers are not permanent, but can be redistrubuted as the branches, which can possess considerable degrees of flexibility, move about. The inclusion of guest molecules within a dendrimer may occur as a result of movements in the branches, which allow temporary inclusion of the guest molecule within the dendrimer structure. In solution, it is assumed that molecules of solvent are able to move with ease through the branches of dendrimers, hopping between such temporary cavities with little or no hindrance. When the solvent is removed, the dendrimer may collapse to a distinctly reduced volume.

As already mentioned, dendrimers resemble conventional polymer molecules in which they adopt differing conformations depending on the thermodynamic quality of the solvent. In all cases, the essential driving force is the need to minimize the free energy of the system. In a good solvent, a dendrimer can adopt an extended conformation, increasing in size to a maximum volume and developing an almost perfectly spherical shape. When the solvent is removed, the dendrimer collapses, and its final shape depends on the overall rigidity of the building components, *i.e.* of the core, the branches and chain-ends.

Free energy considerations also determine the location of the chain-ends, unless the structure is especially rigid, in which case the chain-ends remain at the periphery regardless of the quality of the solvent. For more flexible systems, good solvents cause the chain-ends to remain at the periphery, maximizing interactions between them and the solvent molecules. Less good solvents may encourage chain folding, particularly where the free energy of interaction of chain-ends with branch segments is comparable to or greater than that between chain-ends and solvent molecules. The free energy of interaction may be dominated by the enthalpy term, for example where hydrogen bonding or strong polar interactions can occur. In the absence of a strong enthalpy of interaction, entropy becomes an important factor, and this favours a greater mixing of the structural units, *i.e.* the incorporation of chain-ends within the dissimilar inner chain components.

9.5 Dendrimer Micelles

Dendrimers can be designed to have a hydrophobic interior and a hydrophilic periphery. This gives them properties that are similar to those of conventional surfactants, and they can solubilize hydrophobic substances such as pyridine in aqueous solution by including them as guest molecules. They are therefore effectively *unimolecular micelles*.

Dendrimers can be fabricated so that they have a hydrophobic interior and a hydrophilic periphery. Indeed, organic dendrimers, with relatively non-polar branches naturally have such a structure. As such they can behave like micelles formed by surfactant molecules, and solubilize hydrophobic molecules. They have the advantage that they do not display a critical micelle concentration. Unlike conventional micelles, which are aggregates of small molecules, dendrimers retain their cohesion even at low concentrations, and are effectively unimolecular micelles.

Dendrimer micelles of this type have been formulated as drug delivery vehicles. Dendrimers with a hydrophobic interior have been used to entrap a hydrophobic drug such as indomethacin. This is retained because of the hydrophilic periphery containing ethylene glycol functional groups, and is released slowly because of the collapsed configuration of the interior, through which molecular diffusion is obstructed.

Dendrimers can also be prepared with an inverse relationship between their hydrophobic and hydrophilic constituents, *i.e.* with a hydrophobic periphery and a hydrophilic interior. They can then behave as reverse micelles and are able to concentrate polar molecules from solutions of non-polar solvents. The shape of these molecules, when dissolved in a solvent that matches the hydrophobic nature of the periphery, is spherical with chain-ends extended towards the solvent. The interior may then collapse to a minimum volume, so that unfavourable interactions that might result from penetration by solvent molecules are minimized.

9.6 Characterization of Dendrimers

Dendrimers have been characterized by a variety of general techniques of analytical chemistry, and also by some specialized techniques that are particularly suitable for dendrimers. The following sections describe these techniques as they have been applied to dendrimers. For more details, the reader is referred to the books listed in "Further Reading" section.

9.6.1 Mass Spectroscopy

Mass spectroscopy is a useful technique for the characterization of dendrimers because it can be used to determine relative molar mass. Also, from the fragmentation pattern, the details of the monomer assembly in the branches can be confirmed. A variety of mass spectroscopic techniques have been used for this, including electron impact, fast atom bombardment and matrix-assisted laser desorption ionization (MALDI) mass spectroscopy.

In general, dendrimers show sharp M^+ signals at the expected values. This confirms the overall dendrimer structure. Mass spectra may be complicated by the presence of small peaks corresponding to two- or three-molecule aggregates bearing a single charge and this phenomenon is particularly apparent in negative ion electrospray mass spectroscopy, where the dimer and trimer species aggregate around a single negative charge.

Lower mass/charge ratio peaks may also occur in relatively lower abundance in the mass spectrum. These correspond to fragmentation of the monomer moieties and, in the case of functional dendrimers, can be useful in determining the extent of functionalization. For example, in dendrimers prepared from tertiary butyl esters that have been hydrolysed to carboxylic acid functional groups, the precise number of residual *t*-butyl groups can be determined by mass spectroscopy. This involves a careful study of the fragmentation pattern that arises both from an array of lower molar mass species and from the small amount of unfragmented but charged dendrimer molecule.

9.6.2 Gel Permeation Chromatography

As for conventional linear polymers, gel permeation chromatography (GPC) can be used to find information on the composition of dendrimers, including their polydispersities. Obtaining standards of known relative molar mass and polydispersity is a problem with dendrimers, so the approach that has been taken most often is to use polystyrene standards, as described in Chapter 6.

Using this technique, dendrimers have been found to have extremely low polydispersties, with values typically ranging between 1.02 and 1.04.

9.6.3 High-performance Liquid Chromatography

High-performance liquid chromatography (HPLC) is capable of giving useful information on the homogeneity of dendrimers and dendrons. Normal phase HPLC is generally employed, *i.e.* a relatively non-polar solvent such as hexane or a hexane-rich mixture is used to elute compounds from a polar chromatography column such as coated silica. The advantage of using HPLC is that it can give additional information about compounds that appear, from elemental analysis, to be pure. There are cases where apparently analytically pure dendrons have been shown to contain an impurity, *e.g.* an occasional tertiary butyl ester group replaced by a 2-methylethyl (so-called isopropyl) ester one.

HPLC is limited to relatively low molar mass compounds, *i.e.* below 3000. In practice this tends to restrict the technique to dendrons rather than fully formed dendrimers, but it is nonetheless a useful technique, given the importance of establishing high analytical purity of products at each step in a dendrimer synthesis.

9.6.4 NMR Spectroscopy

Relatively simple NMR techniques have been employed to obtain useful information about dendrimers. For example, the formation of one generation from its immediate predecessor containing NH groups has been followed by ^{1}H NMR. As coupling proceeds, the chemical shifts of the NH protons become shifted to lower field (higher frequency), a shift that is consistent with coupling at the NH groups within the branches.

Spontaneous formation of supramolecular dendrimers from corresponding dendrons has also been demonstrated using ^{1}H NMR. The branching units in dendrons exhibit broad signals as a consequence of the variety of conformations that they can exhibit in solution. As the true dendrimer forms, involving maximum packing of the end groups, the NMR signals become sharp. This occurs because the end groups no longer have the freedom to adopt a variety of conformations, but are held almost completely rigid by the high packing density that prevails.

9.6.5 X-ray Diffraction

X-ray diffraction is a technique that has conventionally been applied to solids to determine the atomic arrangements. It has been used to find out structural information on linear polymers, and has allowed the structure of both crystalline and non-crystalline to be determined. More recently, the technique has been used on dendrimers.

X-ray diffraction patterns from dendrimers tend to lack sharp features and are similar to those from amorphous linear polymers. This suggests a molecular arrangement in dendrimers that is fairly disordered. There is also the problem that these molecules can exist in a large number of energetically equivalent conformations and that in solution there can be rapid interchange between these conformations. This contributes to the overall amorphous structure of dendrimers.

X-ray diffraction does yield some useful information, though, and detailed analysis of diffraction results shows that the monomer branches in dendrimers are arranged in a semi-regular network and that they contain cavities that range in diameter from 0.5 to 1.5 nm. These cavities are uniformly distributed within the dendrimer molecule and are relatively well organized to form channels that connect the surface to the core. The branches in the molecules are not arranged in neat concentric shells, as was first thought, but form a network that has some degree of regularity to it.

9.6.6 Titration

Titration has been used to obtain information about carboxylic acid-terminated dendrimers. In one study of this type of dendrimer, the various generation dendrons were dissolved in aqueous NaOH then precipitated into ethanol. The precipitate was then washed with ethanol to remove any residual NaOH and dried. It was then dissolved in distilled water and titrated with 0.01 M hydrochloric acid, and the pH change recorded.

The dendrimers were found to precipitate between pH 3.1 and pH 4.0. However, the titration curves were not all the same, but varied with the generation of dendrimers. Lower generation species showed a typical s-shaped curve that is characteristic of the behaviour of a weak base titrated with a strong acid. By contrast, with higher generation dendrimers, the end point became progressively less well defined, and in all cases the dendrimer precipitated well before the full equivalent of acid had been added. This latter observation has important consequences for our understanding of dendrimers and may indicate that not all of the carboxylic acid groups could be converted to sodium carboxylate units.

At pH 13, under normal circumstances, all carboxylic acids would be expected to be completely ionized. The difficulty is that, when these groups are relatively closely spaced, as they are at the periphery of a dendrimer, there are significant energetic barriers to achieving full ionization, because each individual ionization event adds more charge to the dendrimer surface. In view of this, it may be that the highest theoretical extent of ionization cannot be obtained, and that the dendrimer does not, after all, become fully neutralized with NaOH. Rather, it seems likely that there is a limit on the extent to which highly charged sodium carboxylate groups can be formed from un-ionized carboxylic acids, and that the observed obscuring of the end point on titration with 0.01 M HCl, as well as the apparently early precipitation of dendrimer is a consequence of this limit. These results suggest that the chemistry of functional groups can be modified significantly by incorporating them into the dendrimers.

9.6.7 Small Angle Neutron Scattering

Small angle neutron scattering (SANS) involves firing neutrons at a sample and measuring the angle through which they are scattered.

This allows the momentum transfer, Q, to be determined according to the equation:

$$Q = \frac{4\pi}{\lambda} \sin \frac{\theta}{2}$$

where θ is the scattering angle. SANS data are typically presented as a plot of intensity against Q.

SANS is a valuable research technique, since it provides information about the size and shape of the molecular scale structures in the size range 1–10 nm found in polymer or dendrimer systems. Neutrons interact with atomic nuclei, hence are sensitive not only to the details of the way such nuclei are arranged but also to the presence of different isotopes.

Facilities for SANS are typically located at national centres, such as the ones at Didcot in the UK or Grenoble in France. Research groups book appropriate amounts of beam time in which to make the experimental measurements.

A variety of types of dendrimer have been studied using SANS and this has provided detailed information on the structures. For example, it has been used to demonstrate that in PAMAM systems the terminal amine groups are dispersed throughout the interior of the dendrimer, rather than remaining at the periphery. It has also shown that the radius of gyration of the dendrimer is proportional to $N^{1/3}$, where N is the number of monomer units in the dendrimer molecule, and that this relationship holds over the entire range of generations. Lastly, SANS has been used to demonstrate that individual dendrimer molecules become more perfectly spherical as the generation number increases.

One way of determining the location of the end groups using SANS is to deuterate them. This also exploits the sensitivity of SANS towards specific nuclei. Using this approach, the structure of amine-terminated poly(propylene imine) dendrimers has been shown to be sensitive to pH. At high pH, where amine groups are un-ionized, the branches have been shown to be extended, resulting in a dense, spherical dendrimer. As pH is lowered, the amine groups become ionized, so there is a structural change to a more hollow, lower density structure.

9.6.8 Atomic Force Microscopy

Atomic force microscopy (AFM) has been used to characterize dendrimers that have been adsorbed onto a surface such as silica. AFM involves moving a finely tipped stylus across a surface and monitoring the tip movements as it traces the surface topography. In studying adsorbed dendrimers, samples can be scanned repeatedly and in a variety of directions. When this is done, it is found that all the images are the same. True dendrimers form objects of only one size.

AFM has also been used to study adsorbed dendrimers of increasing generation number. When this is done, it is found that there is a steady and regular rise in the diameters of the adsorbed dendrimers with each additional generation number, but that each generation of dendrimers remains highly homogeneous.

9.7 Applications of Dendrimers

Dendrimers are versatile chemicals, and their surfaces, interiors and cores can be engineered to provide a range of properties that give them potential applications in several fields of technology. Many of these potential applications aim to exploit the fact that they are highly uniform, can possess multifunctional surfaces and may have internal cavities within them. Examples of possible applications include the following:

(a) *MRI contrast agents*: Dendrimers can be prepared that bind paramagnetic ions and hence can be used for magnetic resonance imaging (MRI) contrast agents. For example, a dendrimer-imaging agent specifically for MRI angiography has been produced and has undergone clinical trials. It is based on polylysine with gadolinium ion complexes on the end groups. These complexes can be varied and one individual dendrimer molecule can include 24 different complexes. This feature allows good signal-to-noise ratios to be obtained. The dendrimer's main function is to prevent the gadolinium complexes diffusing away from the target area, and this produces an MRI picture of excellent contrast and hence greater clinical usefulness.

(b) *Drug delivery*: The possibility of a dendrimer acting as a molecular container for foreign molecules was first suggested by Meijer in 1994. He found that he could trap the relatively small molecules of Bengal Rose dye in the cavities of water-soluble dendrimer molecules. He referred to this as a "dendritic box" and it consisted of a fifth generation poly(propylene imine) consisting of 64 functional groups at the periphery. The small molecules were found to be firmly held in this molecular box, and were unable to diffuse out, unless the system was heated for a prolonged period. Then, the dendrimer box was found to open, and allow the trapped molecules to escape. This suggested the possibility of using the "dendritic box" as vehicle for drug delivery. Meijer's group has been designing a box that could be opened enzymatically or photochemically. Other research groups are also studying this possibility; for example, dendrimers have been considered as controlled release devices for the release of chemotherapeutic agents for cancer treatment. In this approach, a chemotherapeutic drug is weakly bound to the periphery of a dendrimer, which also contains functional groups to increase water solubility. When the dendrimer reaches its target within the body, the bond between the drug and the dendrimer is cleaved enzymatically or photochemically, and the drug is released. Because dendrimers are inert and stable, they are non-toxic to humans and have been shown to be eliminated through the kidneys in the urine.

(c) *Vectors in gene therapy*: Gene therapy is an approach to clinical treatment based on the delivery of therapeutic genes to cells to re-programme them. The therapeutic genes either switch off damaging cell growth or other detrimental activity, or programme the cells to produce their own therapeutic agents. PAMAM dendrimers have been shown to be very effective at the delivery of genetic materials into a variety of cell lines. Both oligonucleotides and DNA have been delivered into cells using these dendrimers. They have several advantages, including the absence of an immune response, low cytotoxicity and the ability to stabilize polynucleic acids against enzymic attack by nucleases.

(d) *Industrial catalysts*: Dendrimers have high surface area and are highly soluble, and this opens up the possibility of using them as nanoscale catalysts. They are effective due to the accessible nature of their active sites, which may lie at the core or at the surface, and they can be readily recovered by ultrafiltration methods. There are two possible catalytic sites in any dendrimer that can be investigated, one at the core and the other at the surface, respectively, and these can be tuned for a variety of technically useful chemical processes. Dendrimers have been found to be effective in enantiomeric catalysis and have been used successfully, for example, in the enantiomeric reduction of benzaldehyde. Many groups worldwide are studying these materials as possible catalysts, and commercial exploitation can be expected in the foreseeable future.

One example has used a manganese porphyrin and iodobenzene encapsulated within a dendrimer to bring about shape-selective epoxidation of alkenes. The important aspect of catalysts is that the reactants can move rapidly to the active site, and that the products can be removed rapidly from the active site and expelled from the dendrimer.

It is possible to take advantage of the differing characteristics of the periphery and the interior to promote chemical reactions. For example, a dendrimer having a non-polar aliphatic periphery with highly polar inner branches can be used to catalyse unimolecular elimination reactions in tertiary alkyl halides in a non-polar aliphatic solvent. This works because the alkyl halide has some polarity, so become relatively concentrated within the polar branches of the dendrimer. This polar medium favours the formation of polar transition states and intermediates, and allows some free alkene to be formed. This, being non-polar, is expelled from the polar region, and moves out of the dendrimer and into the non-polar solvent. This is a highly efficient process, and the elimination reaction can be driven to completion with only 0.01% by mass of a dendrimer in the reaction mixture in the presence of an auxiliary base such as potassium carbonate.

9.8 Megamers

Dendrimers can be assembled into multimolecular assemblies called *megamers*. In some cases, this is because of the natural clustering of dendrimer molecules into supramolecular assemblies. In other cases, it occurs because simple covalent species become incorporated into two or three dendrimers during the process of synthesis.

Synthetic pathways have been deliberately aimed at producing megamers. For example, poly(amido amine) dendrimers of different generations have been combined to give well-defined core–shell megamers in which a central, large core dendrimer is surrounded by a well-defined number of smaller dendrimers.

This is a very new field, but given the current interest in both dendrimer chemistry and supramolecular chemistry, it is one that is likely to receive attention in the future. The prospect of preparing well-defined molecular assemblies, rather than ill-defined clusters, is attractive, and is expected to give access to new properties that can be controlled by the molecular architecture of the assembly. New and useful materials seem likely to emerge as this chemistry grows and is exploited in the years to come.

Further Reading

U. Boas, J. B. Christensen and P. M. H. Heegaard, *Denrimers in Medicine and Biotechnology*, Royal Society of Chemistry, Cambridge, 2006.

F. Vogtle, G. Richardt and N. Werner, *Dendrimer Chemistry*, Wiley-VCH, Berlin, 2009.

D. A. Tomalia, J. B. Christensen and U. Boas, *Dendrimers, Dendrons and Dendritric Polymers*, Cambridge University Press, 2012.

Designing Dendrimers, ed. S. Campagna, P. Ceroni and F. Puntoriero, John Wiley & Sons, New Jersey, USA, 2012.

A. Lederer and W. Burchard, *Hyperbranched Polymers: Macromolecules in between Deterministic Linear Chains and Dendrimer Structures*, Royal Society of Chemistry, Cambridge, 2015.

10 Special Topics in Polymer Chemistry

10.1 Introduction

As with all branches of science, polymer chemistry is continually advancing. New topics in polymer chemistry which involve new concepts, new polymers or novel uses for existing materials are being studied in research laboratories throughout the world. In this chapter, some of the more important of these developments are described including the use of polymers in medicine, electronically conducting polymers, and polymer liquid crystals.

10.2 Polymers in Medicine

Medicine has made major advances in the past 50 or so years partly by the use of devices to improve patient health. These devices include artificial hearts and pacemakers, machines for artificial kidney dialysis, replacement joints for hips, knees, and fingers, and intraocular lenses. These devices need to survive in sustained contact with blood or living tissue.

In order to be successful as part of a medical device a polymer has to resist both biological rejection by the patient's body and degradation. The human body is an environment which is simultaneously hostile and sensitive, so that materials for application in medicine must be carefully selected. The essential requirement is that these materials are 'biocompatible' with the particular part of the body in which they are placed. The extent to which polymers fulfil this

The Chemistry of Polymers, 5th Edition
By John W. Nicholson
© John W. Nicholson 2017
Published by the Royal Society of Chemistry, www.rsc.org

requirement of biocompatibility depends partly on the properties of the polymer and partly on the location in which they are expected to perform. For example the requirements for blood biocompatibility are stringent since blood coagulation may be triggered by a variety of materials. By contrast, the requirements for materials to be used in replacement joints in orthopaedic surgery are less severe and materials as diverse as poly(methyl methacrylate) and stainless steel can be used with minimal adverse reaction from the body.

An idea of the range of materials and applications for polymers in medicine can be gained from the information in Table 10.1. As can be seen from this table a number of polymers are used in medical applications. One particular such polymer is poly(methyl methacrylate), PMMA. Early on it was used as the material for fabricating dentures; later other biomedical applications developed. For example, PMMA is now used as the cement in the majority of hip replacement operations worldwide.

Hip replacement surgery is now routinely used to relieve pain and restore mobility in patients suffering from osteoarthritis. In this condition the surfaces of bone in contact with each other within the joint become worn and the layer of lubricating cartilage disappears. This makes movement of the joint both difficult and painful. By replacing the hip with an artificial joint patients stop experiencing pain and are once again able to move freely.

In a typical hip replacement operation, the top of the thigh bone is removed and a cavity is drilled along the direction of the long axis of the remaining bone. A metal prosthesis is placed in this cavity and secured in place with PMMA cement. In the pelvic girdle a plastic cup is fitted to act as the seat of the new, smaller hip joint. This cup is made of ultra-high molar mass poly(ethylene) and is also secured in

Table 10.1 Polymers and their use in medicine.

Polymer	Use
PVC	Dialysis tubing, catheters, blood bags
Polyurethanes	Pacemakers, blood bags, tubes
Poly(hydroxyethyl methacrylate)	Soft contact lenses, burn dressings
Poly(glycolic acid)	Biodegradable sutures
Nylons	Sutures, haemo-filtration membranes
Poly(methyl methacrylate)	Hard and soft contact lenses, bone cement for artificial joints, intraocular lenses, dentures

place with PMMA cement. The components of an artificial hip joint are shown in Figure 10.1.

Another widely used surgical procedure is knee replacement. In this operation the natural components of the knee joint are replaced with an artificial femoral component made of metal that interfaces with a tibial plateau made of ultra-high molar mass poly(ethylene). Implantation of these joints is becoming more widespread, though the demand for them is not as great as for hip replacement.

Figure 10.1   Components of prosthetic hip joint.

In both procedures, a concern is the extent to which the poly(ethylene) undergoes wear within the artificial joint. The constant rubbing of metal on polymer is capable of generating particles of wear debris, and these can collect around the joint. In extreme cases, this debris can interfere with the metabolic processes in the remaining bone, leading to bone resorption and consequent loosening. Study of wear in these artificial joints and of improvements in the wear characteristics of the poly(ethylene) are now major topics of research in the field of implants.

The PMMA bone cement is formed from a mixture of prepolymer PMMA powder, which contains a free-radical initiator, and liquid MMA monomer. In the operating theatre the powder and liquid are mixed, causing the initiator to dissolve and bring about polymer-isation in the monomer component. The powder pre-polymer does not dissolve in the monomer but remains in the newly polymerised materials as a kind of reinforcing filler.

Once polymerisation is complete, the components of the new hip joint can be connected together and the operation completed. This surgical procedure has been very successful over the past 30 or so years and now an estimated 45 000 such operations are carried out each year in the UK alone. Similar procedures are used for the re-placement of both arthritic knees and arthritic fingers, though these latter operations are less common than hip replacements. Overall considerable amounts of PMMA are used each year as bone cements for these surgical procedures.

10.3 Ionomers

Like the word nylon the term 'ionomer' was invented by the Du Pont company in America to cover a class of polymeric material, consisting of an organic backbone bearing a small proportion of ionisable functional groups. The organic backbones are typically hydrocarbon or fluorocarbon polymers and the ionisable functional groups are generally carboxylic or sulfonic acid groups. These functional groups, which reside on no more than about 10% of the monomer units in the polymer, may be neutralised, for example with sodium or zinc ions.

The presence of these ionic groups gives the polymer greater mech-anical strength and chemical resistance than it might otherwise have. Indeed the ionomer is resistant to dissolution in many solvents because of its unconventional chemical character, often being too ionic to dis-solve in non-polar solvents and too organic to dissolve in polar solvents.

A variety of ionomers have been described in the research literature, including copolymers of (a) styrene with acrylic acid, (b) ethyl acrylate with methacrylic acid, and (c) ethylene with methacrylic acid. A relatively recent development has been that of fluorinated sulfonate ionomers known as Nafions, a trade name of the Du Pont company. These ionomers have the general structure illustrated (10.1) and are used commercially as membranes. These ionomers are made by co-polymerisation of the hydrocarbon or fluorocarbon monomers with minor amounts of the appropriate acid or ester. Copolymerisation is followed by either neutralisation or hydrolysis with a base, a process that may be carried out either in solution or in the melt.

$$(CF_2CF_2)_n \, (CF_2CF)_y$$
$$|$$
$$O-CF_2CF(CF_3)OCF_2CF_2SO_3^-M^+$$

(10.1)

The presence of ions in an otherwise organic matrix is not thermodynamically stable. As a result these materials undergo slight phase separation in which the ions cluster together in aggregates. These ionic clusters are quite stable and may contain several water molecules around each metal ion. They act partly as crosslinks and partly as reinforcing filler, which is the reason for the greater mechanical strength that ionomers show.

Ionomers are able to form crystalline domains, and when they do so, they typically show unusual crystallinity. On cooling from the melt, they develop relatively thick crystals, but crystallisation continues at room temperature and results in the formation of thinner secondary crystals over periods of several days. This means that the properties of ionomers change over time, as this growing crystallinity leads to room-temperature annealing. It also leads to unusual melting behaviour, because on heating there are two distinct melting events, with the secondary crystals melting first.

Ionomers are used to prepare membranes for a variety of applications including dialysis, reverse osmosis, and in electrolytic cells for the chlor-alkali industry. This latter application needs materials that show good chemical resistance and ionomers based on per-fluorinated backbones with minor amounts of sulfonic or carboxylic acids are ideal. They also show good ion-exchange properties.

The term *chlor-alkali* refers to those products obtained from the commercial electrolysis of aqueous sodium chloride. These are chlorine, sodium hydroxide, and sodium carbonate. The first two are produced simultaneously during the electrolysis while the latter is included because it is also produced in small quantities and shares many of the end uses of sodium hydroxide. Perfluorinated ionomer membranes are permeable to sodium ions but not the chloride ions, and hence they are useful for these electrolytic cells. The arrangement of a typical membrane cell is shown in Figure 10.2.

Ionomer membranes show good ion selectivity. They are able to distinguish between ions on the basis of size and charge, and show such good selectivity that they have also been used for membranes in experimental ion-selective electrodes. Their main use, though, remains in membrane cells of which numerous examples are currently employed throughout the world's chlor-alkali industry (Figure 10.3).

Figure 10.2 Arrangement of a typical electrolytic chlor-alkali cell.

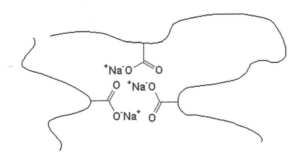

Figure 10.3 Structure of a typical ionomer.

10.4   Electronically Conducting Polymers

A field of research that has been developing in recent years is that of conducting polymers. Conventional organic polymers, such as poly(ethylene), are such poor conductors of electricity that they are widely used as insulators. However, a number of polymers are now known which have highly conjugated structures, including poly(acetylene) and poly(pyrrole). Consequently they can be doped to give materials which exhibit electrical conductivities approaching those of metals; for this reason these materials are sometimes called synthetic metals. Poly(p-phenylene) can also be doped to give high conductivity and this discovery in 1979 led to the development of a number of conducting polymer systems that were based on molecules containing aromatic rings.

The conductivity of these polymers is due to the presence of conjugated p-orbitals on the backbone carbon atoms. These p-orbitals are orthogonal to the three bonding orbitals of the carbon atoms, which are all sp^2 hybridised, and hence lie in a single plane at angles of approximately 120° to each other. The p-orbitals overlap sufficiently to form a delocalised set of orbitals that cover the whole polymer molecule. They thus form a one-dimensional band of energy levels in which the electrons can become mobile when the polymer is doped with an oxidising species. Oxidation removes some of the electrons from the delocalised band, leaving the remaining electrons free to move within the band and across the molecule when subject to a potential difference.

A wide range of substances have been used as dopants in these systems, including AsF_5, I_2, SbF_5, $AlCl_3$, $ZrCl_4$, IF_5, $MoCl_5$ and WCl_5. They are all good oxidising agents and their presence confers reasonable conductivity on the resulting polymer system. Conductivities are not as high as for true metals (see Table 10.2), though at the upper end of the range, they approach values usually associated with metals and thereby justify the term *synthetic metal* for these systems.

Table 10.2   Conductivity values for various materials.

Material	Conductivity/S cm^{-1}
Copper	8×10^5
Iron	10^5
Mercury	10^4
Poly(acetylene)	10^{-6} to 10^3
Poly(p-phenylene)	10^{-15} to 5×10^2

Most conducting polymers, such as doped poly(acetylene), poly(*p*-phenylene), and poly(*p*-phenylene sulfide), are not stable in air. Their electrical conductivity degrades rapidly, apparently due to reaction with oxygen and/or water. Poly(pyrrole) by contrast appears to be stable in the doped conductive state.

Polymers for these conductive systems may be synthesised by a variety of means including Ziegler–Natta polymerisation or nucleophilic displacement reactions. The molecules tend to be rigid because of the need for them to possess extended conjugation. This lack of free rotation about carbon–carbon bonds within the molecule imposes a high energy barrier to solvation, thus making these molecules difficult to dissolve. This lack of solubility in turn makes it very difficult to characterise these polymers so that molar mass or degree of polymerisation tend not to be known with any degree of accuracy for a given conductive polymer.

There are difficulties in analysing conductive polymers, and information on the relationship between structure and properties somewhat difficult to obtain. These materials have already found a variety of uses, including flat panel displays, antistatic packaging and rechargeable batteries, and other applications are likely to emerge in the future.

Although the polymer component is almost always conjugated for synthetic metals this requirement is not absolute. In 1988 M. K. Thakur at AT&T Bell Laboratories prepared conducting polymer complexes from natural rubber doped with iodine. The doped rubber has a conductivity some ten orders of magnitude greater than that of the undoped material. The conductivity in this system appears to be primarily electronic with charge carriers hopping between sites on different polymer chains. Other polymers also become conducting when doped with iodine, including the *trans*-isomer of poly(isoprene), but unsubstituted poly(butadiene) does not. This indicates the importance of electron-releasing groups in establishing the electrical conductivity of this system.

Finally, synthetic metals made of polymeric organic molecules may also show the property of ferromagnetism. Organic materials of this kind were first demonstrated in 1987 by Ovchinnikov and his co-workers at the Institute of Chemical Physics in Moscow. The polymer they used was based on a polydiacetylene backbone, which contains alternating double–single and triple–single bonds between the carbon atoms of the molecule (10.2).

$$(-C\equiv C-C=C-C\equiv C-)_n$$

(10.2)

This highly conjugated molecule was stabilised with nitroxyl biradical side chains. The resulting material had sufficient ferromagnetism that a usable compass needle could be made from it. Despite the success of this demonstration, organic ferromagnetism remains a curiosity. Such polymers are not likely to replace conventional ferromagnetic metals in any application within the foreseeable future.

Conductive polymers are difficult to process and this has limited their applications so far. They have been used as antistatic coatings and in commercial displays and batteries. Other applications that are currently the subject of research include organic solar cells, light-emitting diodes and biosensors. However, none of these areas has yet resulted in commercial products.

10.5   Interpenetrating Polymer Networks

An interpenetrating polymer network (IPN) is defined as a material comprising two or more networks which are at least partly interlaced on a molecular scale, but not covalently bonded to each other. These networks cannot be separated unless chemical bonds are broken. Two possible methods exist for preparing them, as follows:

(i) A two-step process involving formation of the first network, which is then swollen by a solvent containing the monomer of the second network, and the second monomer allowed to polymerise to form the second network;

(ii) A one-step process involving simultaneous synthesis, where the monomers of both networks are mixed, then allowed to polymerise.

An IPN has different properties from either a copolymer or a polymer blend. It may swell in solvents, but will not dissolve; it will resist creep or flow to a greater extent than copolymers or blends. Some differences in the physical properties of IPNs compared with polymer blends can be seen in Table 10.3. The major reason for the differences in properties between polymer blends and IPNs is that the latter have greater adhesion and better mixing.

The interconnection between the two different polymers can cause some essentially thermoplastic polymers to behave as thermosets

Table 10.3 Properties of polyurethane/acrylic IPNs compared with blends and homopolymers.

Polymer	Tensile strength/Mpa	Elongation at break/%
IPN 80/20	49.0	780
IPN 60/40	45.0	540
Linear blend 80/20	12.6	640
Linear blend 60/40	24.9	357
Homopolymer polyurethane	42.1	640
Homopolymer polyacrylate	17.7	15

when combined in an IPN. This arises because of chain entanglement between different polymers. The entanglements behave as crosslinks at lower temperatures but they can be broken by heat.

The term IPN was first used in 1960 to describe the apparently homogeneous product obtained from styrene crosslinked with divinylbenzene. IPNs were prepared from this system by taking a crosslinked poly(styrene) network and allowing it to absorb a controlled amount of styrene and a 50% divinylbenzene–toluene solution containing initiator. Polymerisation of this latter component led to the formation of an IPN, the density of which was greater than that of the original polymer. These materials were used as models for ion-exchange resins.

Since this pioneering work a number of IPNs have been prepared. Poly(styrene) has been used as the second network polymer in conjunction with several other polymers, including poly(ethyl acrylate), poly(*n*-butyl acrylate), styrene–butadiene, and castor oil. Polyurethanes have been used to form IPNs with poly(methyl methacrylate), other acrylic polymers, and with epoxy resins.

Typically IPNs exhibit some degree of phase separation in their structure depending on how miscible the component polymers are. However, because the networks are interconnected such phase separation can occur only to a limited extent, particularly by comparison with conventional polymer blends. Polymer blends necessarily have to be prepared from thermoplastics; IPNs may include thermosets in their formulation.

IPNs are found in many applications though this is not always recognised. For example conventional crosslinked polyester resins, where the polyester is unsaturated and crosslinks are formed by copolymerisation with styrene, is a material which falls within the definition of an interpenetrating polymer network. Experimental polymers for use as surface coatings have also been prepared from IPNs, such as epoxy–urethane–acrylic networks, and have been found to have promising properties.

10.6 Inorganic Polymers

A number of polymers exist which contain inorganic backbones but which carry organic groups attached to the backbone. The best known examples of such polymers are the silicones which, as was described in Chapter 1, consist of alternating silicon and oxygen atoms along the backbone, with either Si—O—Si crosslinks or organic groups attached to the silicon atoms.

Poly(phosphazenes) are similar, partly inorganic polymers in that they consist of inorganic backbone, in this case of nitrogen and phosphorus atoms. They are separated formally by alternating single and double bonds and carry organic groups on the phosphorus atoms (10.3).

$$[-N{=}P-]_n$$
with R above and R below the P

(10.3)

The poly(phosphazene) backbone can be made to be part of a fully inorganic polymer by replacing the organic substituents with chlorines, thus generating poly(dichlorophosphazene). However, this is a reactive molecule and is more usually employed as the starting material for the preparation of the various partially organic poly(phosphazenes).

Among other uses, these polymers have been employed in a variety of biomedical applications. Poly(phosphazenes) containing organic side chains, derived from the anaesthetics procaine and benzocaine, have been used to prolong the anaesthetic effect of their precursor drugs. They have also been used as the bioerodable matrix for the controlled delivery of drugs.

Another group of inorganic polymers investigated in the last ten years or so are the polystannanes, *i.e.* polymers based on the repeat unit:

$$*{-}[Sn]_n{-}*$$
with R above and R below the Sn

These substances can be synthesised from dialkyltin dihydrides, R_2SnH_2, by reaction in the presence of $RhCl(PPh_3)_3$ (Ph = phenyl), so-called Wilkinson's catalyst. The process is termed dehydropolymerisation, and results in yellow materials with number average molar masses in the range 1000–70 000, and a polydispersity of 2–3. The polymer backbone consists of tin atoms only, and is relatively stiff. Consequently, these materials show natural semi-conductor properties and can be rendered conductive by doping with SbF_5. They also show liquid crystal behaviour.

As well as polystannanes, the analogous germanium polymers, polygermanes, have also been prepared. These are only oligomeric, and so far have not been prepared with more than five germanium atoms in the chain. They have not been studied as extensively as the polystannanes, and less is known about their properties.

From the preceding discussion it appears that purely inorganic polymers are rarely encountered. However, this is not necessarily true. If we assume that what we are looking for is an inorganic material containing directional covalent bonds making up a large chain structure, there are in fact a number of common materials which fulfil the requirements; these include the silicate minerals and silicate glasses.

The concept of silicates as inorganic polymers was implicit in the ideas developed by W. H. Zacheriasen in the early 1930s. He conceived of silicates as consisting of macromolecular structures held together by covalent bonds but including network-dwelling cations. These cations were not assumed to have a structural role but merely to be present in order to balance the charges on the anionic polymer network.

Modern theories of silicate structure do not assign the cations such a passive role, but assume that they are site bound in well-defined regions around the macromolecule. This is essentially similar to co-ordinating ions with ionic organic polymers, for example in the calcium or zinc salts of poly(acrylic acid). However, the fundamental concept of Zacheriasen remains: these silicate minerals and glasses, in being made up of covalently linked silicon–oxygen structures, are essentially macromolecular. Silicate minerals and glasses are thus among the true inorganic polymers.

The nature of the polymer backbone in inorganic macro-molecules is very different from that of organic polymers. For inorganic molecules, —Si—Si— bonds are not sufficiently stable to form the basis of a polymeric structure. Instead, as we have seen, the backbone comprises alternating silicon–oxygen atoms. Such oxy-bonds are polar.

This means that chain termination is generally achieved by forming a charged group. The presence of divalent cations to preserve electrical neutrality with such a macromolecular system tends to result in ionic crosslinks. From this it follows that inorganic polymers are always highly crosslinked, this crosslinking being either covalent or ionic in character.

The role of oxygen in these inorganic polymers in important. In Zacheriasen's theory, oxygen was described as *bridging* if part of a covalent structure (10.4). When there was an ionic bond, the oxygen was described as *non-bridging* (10.5).

$$-\overset{|}{\underset{|}{Si}}-O-\overset{|}{\underset{|}{Si}}- \qquad\qquad -Si-O^- \;\dots\; M^+$$

$$(10.4) \qquad\qquad\qquad (10.5)$$

A typical inorganic polymer is found in the pyroxene mineral wollastonite. This is simply a calcium silicate of empirical formula $CaSiO_3$. The siloxane backbone is predominantly linear (10.6). This highly charged chain is neutralised by an appropriate number of calcium ions which act as ionic crosslinking agents between the chains. Thus wollastonite is characterised by being very brittle and having a high melting point.

$$\begin{array}{ccc} O^- & O^- & O^- \\ | & | & | \\ -Si-O-Si-O-Si-O \\ | & | & | \\ O^- & O^- & O^- \end{array}$$

$$(10.6)$$

A variety of other structures are possible with silicate minerals, including sheets and three-dimensional frameworks. In all cases, the structure includes bonds that are predominantly covalent and directional. They can therefore be viewed as being based on increasingly crosslinked inorganic polymers.

Silicate glasses are similarly considered to be based on inorganic polymer molecules. Zacheriasen did not state this idea implicitly in his work; indeed his ideas did not derive from polymer chemistry at all but from fundamental studies in crystal chemistry. He regarded the structure of glasses as extended random networks of linked silicon–oxygen tetrahedra, these tetrahedra being linked with each other at the corners. As in the silicate minerals these siloxane macromolecules are negatively charged and are associated with network-dwelling cations whose primary role is to balance the charge to achieve electrical neutrality. The main difference between these glasses and the silicate materials is that the glasses, being super-cooled liquids and not true solids, have a much less orderly structure. Despite this, they too are now recognised as being examples of inorganic polymers and present studies based on the concepts of polymer chemistry are increasing our understanding of the nature and properties of this important group of materials.

10.7 Polymer Liquid Crystals

Low molecular mass compounds capable of forming liquid crystals have been known since the late 1880s. They did not assume commercial importance until the late 1960s, however, when their properties were exploited in the design of electronic displays. Following the development of commercial applications for liquid crystals, polymers began to be studied for their potential in this application.

The liquid crystalline phase is called a mesophase and is intermediate between solid and liquid. In this mesophase the molecules show liquid-like long-range behaviour, *i.e.* are essentially disordered, but also some crystal-like aspects of short-range order. The type of long-range order in the mesophase may vary. If the molecules align themselves as layers this is described as a smectic phase; if the alignment is as parallel threads it is described as a nematic phase.

Typical materials that exhibit liquid crystalline behaviour are made up of long, thin molecules. Hence in principle polymers ought to show the basic requirement for liquid crystal behaviour. Conventional polymers, however, are too flexible and tend to adopt random coil configurations in the melt. They are thus not sufficiently anisotropic to exhibit a mesophase.

In order to make polymers behave as liquid crystals it is necessary to introduce some structural rigidity. A typical polymer which has the required rigidity is poly(phenylenetetraphthalamide) (10.7). This material belongs to a class of polymer known as the aramids. Other liquid crystalline polymers are the thermotropic polyesters derived from *p*-hydroxybenzoic acid, *p,p'*-biphenol and terephthalic acid (10.8).

$$-(NH-\!\!\!\bigcirc\!\!\!-NH-\underset{\underset{O}{\|}}{C}-\!\!\!\bigcirc\!\!\!-\underset{\underset{O}{\|}}{C}-)_n$$

(10.7)

(10.8)

Unlike low molar mass liquid crystals, these materials do not undergo a nematic–isotropic transition. Instead, they adopt liquid crystal behaviour throughout the region of the phase diagram for which they are in the melt. Above a particular temperature, rather than adopting an isotropic liquid structure, they decompose.

Certain polymers will exhibit liquid crystal behaviour in solution. This phenomenon was first noted for natural polypeptides, but was found to be obtainable in synthetic systems initially for poly(γ-benzyl L-glutamate) (10.9) in chloroform.

(10.9)

para-Linked aromatic polyamides will also show liquid crystal character in appropriate solvents. Rod-like aromatic polyamides are soluble in solvents such as dimethylacetamide, N-methylpyrrolidone, and tetramethylurea containing lithium or calcium chlorides. They are also soluble in concentrated sulfuric acid, oleum, and hydrogen fluoride. The precise conditions under which these polymers will form liquid crystalline solutions varies with molar mass, polymer structure, and temperature. This liquid crystalline behaviour is exploited technically in the preparation of highly oriented aramid fibres which show good strength in the direction of the orientation.

Since the earliest discoveries of polymeric liquid crystalline melts and solutions a large number of such systems have been reported and this continues to be a vigorous field of research.

Liquid crystal polymers are chemically inert, and resist attack by most chemicals, even at elevated temperatures. They thus resist degradation by organic solvents, strong acids and bases and high-temperature steam. Polar liquid crystal polymers are capable of forming high strength fibres, as illustrated by the material kevlar, which is a polymer liquid crystal used to make helmets and bullet-proof vests, as well as being used in other applications where the combination of light weight and strength are needed. This includes being used in mechanical components, food containers and other specialised packaging systems.

Liquid crystal polymers can also be used to store information. The liquid crystal melt is held between two conductive glass plates and the side chains are oriented by an electric field to produce a transparent film. The electric field is turned off and the information inscribed on to the film using a laser. The laser has the effect of heating selected areas of the film above the nematic–isotropic transition temperature. These areas thus become isotropic and scatter light when the film is viewed. Such images remain stable below the glass transition temperature of the polymer.

Like the chemistry of the materials, there is much research activity on the topic of applications for liquid crystals. So far a number of commercial products are available which exploit these devices but such is their technological promise that numerous further developments are predicted for the future.

10.8 Polymers for Food Packaging

Both thermosets and thermoplastics are used as food-contact materials, though thermoplastics predominate in this application. Examples of the former are phenol– and urea–formaldehyde, while probably the best known example of the latter is low-density poly(ethylene). Other linear polymers are used include high-density poly(ethylene), poly(propylene), and PVC, all of which find quite extensive use. Polymers for food packaging may be in the form of films and other flexible items, or in the form of rigid containers, such as clear drinks bottles or opaque cartons for dairy products.

For packaging, there are a variety of requirements. The foodstuff must be preserved for the customer until the moment of consumption in a form that is palatable and wholesome. This means the polymer will usually require good barrier properties, both to prevent loss of moisture from the food, and to prevent ingress of undesirable, possibly odorous contaminants from outside. In order to obtain optimum preservative properties, polymers are often employed in multilayers, each layer being a different polymer, so that the whole polymer system presents a barrier that is appropriately impervious. Unfortunately, blended films and used containers of this kind are difficult to recycle, and thus present a potential environmental problem (see Chapter 11).

It is critical that plastics used in food contact applications are not themselves a source of contamination of the food. High molar mass polymers, because of their limited solubility in either organic or aqueous media, rarely if ever migrate into the food and hence present little toxic hazard. However, in order that polymers have the necessary properties of stability towards heat and light, as well as storage stability and flexibility, a variety of additives are blended into the polymer, and these do have the potential to migrate into the food, causing problems ranging from alteration of the taste to possible toxicity.

In a typical polymer used for packaging there may be a number of low molar mass substances capable of migrating on contact with particular foodstuffs. These fall into one of two categories:

(i) Polymerisation residues, *i.e.* free monomers, catalysts and solvents; or
(ii) Processing aids and additives, *e.g.* antioxidants, stabilisers, lubricants, fillers and fungicides.

There is a considerable volume of legislation throughout the world governing what may be allowed to migrate from a polymeric packaging material into food in contact with it. Typically, within European Union legislation, no more than 10 mg dm^{-3} of any component may be allowed to migrate into the food, a figure which is modified to 60 mg kg^{-1} of foodstuff for those items whose contact area cannot be readily measured.

Legislation limits not only the amount, but also the nature of what may be allowed to migrate. For example, the main plasticisers allowed for food contact polymers are listed in Table 10.4.

Table 10.4 Permitted plasticisers in food-contact polymers.

Polymer	Plasticiser
PVC film	Di-(2-ethylhexyl) adipate
	Polyesters of dicarboxylic acids and diols, average molar mass 2000
Vinylidene dichloride film	Epoxidised soya bean oil
	Acetyl tributyl citrate
PVC closure seals for jars	Di-(2-ethylhexyl) phthalate
	Di-iso-decyl phthalate
	Di-iso-octyl phthalate

Analysis of substances migrating from food contact plastics is possible at very low levels in real foods. Volatile species are the easiest to determine, since the straightforward approach of 'head-space' analysis can be employed. This technique analyses the gaseous space above a sample in sealed vial that has been allowed to equilibrate. Analysis of a fixed volume of this headspace vapour is withdrawn and its composition determined, typically by capillary gas chromatography or high-performance liquid chromatography. Such analysis is critical in establishing the nature and level of species migrating from any particular packaging plastic, and thus of informing the regulatory process.

In the future it is anticipated that requirements for polymeric packaging will change. One trend driving this change is the increasing use of microwave ovens to prepare individual portions of food. Another is people's increasing habit of consuming frequent snacks, and the decline of the 'family meal'. However undesirable these latter trends, they are clear and will influence the way polymers are used to package and preserve food. Regulatory control, too, will change and will increasingly require the industry to supply data concerning the toxicity of both individual components and polymer formulations used in food-contact packaging. This concern, together with increased public awareness of pollution from discarded packaging materials, may well influence the use of polymers in food-contact. However, polymers have such desirable qualities in preserving food in the freshest possible state as it is transported from producer to consumer that we can be confident that they will go on being used for this purpose well into the future.

10.9 Polymers for Controlled Drug Delivery

So-called controlled drug delivery is an important topic in contemporary human health, and is a field in which polymers find a variety of applications. This field is growing rapidly and the study of polymers for drug delivery is a major topic of research worldwide.

Delivery of drugs in a controlled way has several advantages over conventional delivery, either by oral dose or by injection. It allows drugs that are otherwise metabolised rapidly to be delivered over clinically useful time periods. It also allows the concentration of a drug within the body to be maintained at the therapeutic level but below the level at which adverse side-effects occur. The latter effect contrasts significantly from the levels of drug in the system with conventional delivery. Here, the amount of active compound varies widely, with a surge immediately after dosing, followed by a decline to levels at which there is minimal or even no therapeutic effect. This results in a pulsed delivery of drug, the effectiveness of which is dependent on the patient co-operating by taking medication at the appropriate times.

If a drug is delivered as part of a polymeric delivery system, the polymer employed must be capable of being removed from the body by some sort of natural process. This means that either the polymer or its degradation products must be able to be excreted from the body *via* the kidneys.

Several different polymers are employed for drug delivery, and the reader is directed to the references at the end of this section for more comprehensive accounts of these polymers and the systems in which they have been employed. However, we can briefly observe that typically bespoke copolymers are used for this purpose, with the monomer components being selected to impart specific attributes. These include the rate of degradation, hydrophobicity, and compatibility with the active drug molecule. Examples of monomers used in this way are given in Table 10.5.

Some of these monomers are especially important in polymers for controlled drug delivery. These include lactic acid (both D and

Table 10.5   Segments used in drug delivery polymers.

Segment	Comments
Poly(ethylene)	Allows prolonged release
Poly(propylene)	Often used in ophthalmic application
PVC	Used in membranes
Poly(vinyl alcohol)	Surface stabiliser for microspheres. Also used in bio-adhesive hydrogels
Poly(vinyl acetate)	Enhances drug permeability
Poly(acrylic acid)	Bio-adhesive polymer. Hydrogels swell with change in pH
Poly(N-isopropyl acrylamide)	Used in heat-sensitive microgels
Poly(methyl methacrylate)	Bone cement able to release active molecules
Poly(ethylene glycol)	Numerous applications in drug delivery
Poly(lactic acid)	Degradable species used either as homopolymers or copolymers to deliver drugs by degradation of carrier matrix
Poly(glycolic acid)	
Polyurethanes	Wide use for time-dependent release

L forms) and glycollic acid, polymers and copolymers of which degrade readily in aqueous media. Of these, the homopolymer of L-lactic acid, poly(L-lactic acid), PLLA, is the most biocompatibile because the monomer is so readily tolerated by the tissues of the body. It is also readily excreted by the kidneys. Both possible optical isomers of lactic acid can be used to prepare useful polymers for drug delivery, and they can also be combined with glycollic acid and/or other monomers to produce versatile polymers that can be degraded in the body.

Poly(ethylene glycol), PEG, is also widely used to prepare controlled release polymers. These are typically block copolymers, containing substantial blocks of PEG in molecules containing blocks of other monomer species, such as lactic acid.

Drug release from these polymeric systems may occur by a number of mechanisms. With a completely insoluble polymer, the drug may be eluted *via* a diffusion mechanism. In the case of degradable polymers, the drug is gradually released as the polymer matrix is degraded. Polymers with charged functional groups, such as acrylic acid copolymers, typically change their dimensions with pH, and this makes them sensitive to changes in local pH, so that changing the external pH may cause a change in the polymer conformation that effectively squeezes the drug out into the surrounding tissues.

Polymer-drug conjugates can be injected into the bloodstream, where they circulate, until delivering the drug at the designated treatment site. This approach is particularly under investigation for treatment of tumours in cancer therapy.

Polymer morphology is also important in these applications, and polymeric drug delivery has been achieved using a variety of morphologies, including microspheres, micelles and hydrogel matrices. The nature of the polymer/copolymer and its morphology can influence the specificity of the drug targeting, improve absorption rates and protect active drugs from premature elution and decomposition. Selection is therefore a complicated process, but offers the potential for the production of device that can release precise amounts of active drug exactly at the target location, and to sustain this beneficial level of release for a prolonged period of time.

Internal delivery of drugs from polymeric controlled release devices is not the only application for polymers in pharmacy. They can also be used in transdermal delivery devices. These are devices that promote the passage of the drug across the skin, possibly in association with chemicals designed to enhance skin penetration. This approach to delivery is being used for an increasing number of medicines in modern pharmacy.

Polymers are important components of transdermal delivery devices. The devices typically consist of several layers, and may employ a number of different polymers. They may contain the drug either in a small reservoir or as a drug-polymer matrix sandwiched between two polymeric layers.

The polymers used for transdermal patches are similar to those used for internal devices. For example, crosslinked PEGs and acrylic acid copolymers are used. Semi-synthetic polymers, such as hydroxypropyl methylcellulose and ethyl cellulose are also employed. Patches must adhere to the skin surface and this is typically achieved by formulating pressure-sensitive adhesives based on acrylic systems, poly(*iso*-butylene) or silicones. The choice of adhesive is based on a variety of considerations, including adhesion properties and skin compatibility. Full details can be found in the references.

10.10   Lithium–Polymer Batteries

The term *lithium polymer battery* is ambiguous and can be applied to two different kinds of system. It is most generally applied to commercial batteries that use electrochemistry based on lithium ion transport within generally a gelled liquid, all housed in a non-rigid polymer pouch or container. However, the term can also be applied to electrochemical cells involving lithium ion transport within true polymer electrolytes. To date, the latter have not been commercialised, though they have been subjected to a considerable amount of research.

The commercial items known as lithium polymer batteries (or cells) first appeared on the market in 1991. They contain the four essential components of a cell (*i.e.* positive and negative electrodes, electrolyte and separator) in forms that may include a polymer component. In addition, they are held in a light, flexible polymer casing. The positive electrode typically comprises a complex oxide of lithium with a transition metal, such as $LiCoO_2$ or $LiFePO_4$, a conductive additive and a binder made of poly(vinyldene fluoride). The negative electrode contains similar components, although with carbon (graphite) in place of the lithium compound.

Lithium ion batteries of this type work as follows: At the charging stage, lithium ions are driven out of the metal oxide/phosphate electrode and move through the electrolyte solution to the graphite-based electrode, where they settle between the layers of carbon atoms within the graphite structure. In this way, the battery takes up energy and stores it.

When the battery is used to provide power, the lithium ions move back out of the graphite electrode, through the electrolyte and return to the metal oxide electrode. This causes the graphite electrode to be negative and the metal oxide electrode to be positive, and produces the energy that the battery uses to drive an external device. Movement of Li^+ ions between the electrodes within the electrolyte causes electrons to flow around the outer part of the circuit in the opposite direction. Electrons themselves do not flow within the electrolyte; only lithium ions move within this medium. Movement of electrons is restricted to the external circuit only.

Lithium polymer batteries of this type have several advantages. They are light and can be fabricated into a variety of shapes. They are therefore suitable for use in modern, compact devices such as mobile phones, laptops and portable media players. They are also being investigated as batteries to power electric vehicles.

As we have seen, lithium polymer cells of the alternative type, *i.e.* using polymer electrolytes, are being widely investigated. Solid polymer electrolytes based on lithium bis(fluorosulfonyl)imide blended with poly(ethylene oxide) have been studied, as have electrolytes based on ionic liquids absorbed into microporous blends of vinylidene fluoride–hexafluoropropylene copolymer with poly(methyl methacrylate). Prototypes of these batteries have been prepared but so far none has made it to the marketplace.

10.11   Co-ordination Polymers

The term *co-ordination polymer* was originally applied to polymers made using Ziegler–Natta catalysts, since this process involves the transient formation of co-ordination complexes during the polymerization process. However, the term is increasingly applied to macromolecules that are formed due to co-ordination of multi-dentate ligands to metal ions. These polymers form by having co-ordination complexes as their repeat units.

In typical co-ordination polymers, a ligand formally behaves as a Lewis base by donating a lone pair of electrons to a metal ion (Lewis acid) to form a dative covalent bond. If the ligand is multi-dentate, it can co-ordinate to more than one metal ion. Consequently, a macromolecular assembly can be built up using the appropriate quantities of ligand and metal ion (see the example in structure 10.10). Mono-dentate ligands may also be included in the mixture, and act as terminal groups because they do not provide a site for further building of the network.

(10.10)

The metal ion employed can influence the nature of the final assembly, and determine whether the co-ordination polymer has a one-, two- or three-dimensional structure. Ligands can also influence the nature of the structure, depending on their flexibility. Highly flexible ligands with freedom to rotate about bonds and to alter their orientation within a structure are able to create considerable variety within the assembly. Ligands commonly co-ordinate through oxygen and nitrogen atoms, though sulfur and phosphorus can also be used. Typical substances which can act as ligands include polypyridines, phenanthrolines, hydroxyquinolines and polycarboxylates. Metal ions of all kinds can be used, and co-ordination polymers containing complexes of a variety of transition and main group metals have been reported in recent years.

Co-ordination polymers are not yet widely used in any commercial applications. However, they have interesting properties that could be exploited in devices such as sensors. For example, the co-ordination polymer formed by cobalt ions with $[Re_6S_8(CN)_6]^{4-}$ clusters are orange when they include water in their structure. Replacement of this water by either THF or diethyl ether causes a distinct colour change to purple, green or blue, depending on the details of the co-ordination polymer complex. This colour change is caused by a change in the co-ordination number of cobalt ions from six to four and provides a way fro the complex to act as a sensor for these particular solvents.

Further Reading

C. Alvavarez-Lorenzo and A. Concheiro, *Smart Materials for Drug Delivery*, Royal Society of Chemistry, Cambridge, 2013.

A. Misra and A. Shahiwala, *Applications of Polymers in Drug Delivery*, *Smithers RAPRA Technology*, Shawbury, Shrewsbury, 2014.

R. A. Bader and D. A. Putnam, *Engineering Polymer Systems for Improved Drug Delivery*, Wiley, Chichester, 2014.

11 Polymers and the Environment

11.1 Introduction

The impact of the industrialised nations on the world environment is a topic of growing concern. Consumption of all kinds of raw materials continues to grow inexorably and this is associated with widespread and indiscriminate dumping of the residues of these materials as waste products of various kinds. Plastics are a significant part of this waste, and their petrochemical origin, as well as the energy used in their production, make a significant contribution to the world's pollution problems.

This chapter looks at the particular question of polymers and the environment. A book on the science of polymers ought to give some consideration to this issue because of the growing concern about the effect of discarded plastic on the quality of life on Earth. If we are to produce and use polymers in ways that are more environmentally responsible (and society will increasingly demand that we do) then there will be new technical challenges to face. One purpose of this chapter is to indicate what some of these challenges might be.

11.2 Pollution by Polymers

Pollution is defined as the introduction by man into the environment of substances liable to cause hazards to human health, harm living resources and ecological systems, damage structure or amenity, or interfere with legitimate use of the environment. This all-embracing

The Chemistry of Polymers, 5th Edition
By John W. Nicholson
© John W. Nicholson 2017
Published by the Royal Society of Chemistry, www.rsc.org

definition includes the impact of polymers on the environment, both in their original form as raw materials, for example in the form of thermoplastic pellets, and in their finished form, as discarded items fabricated for a particular use. In both forms, plastic waste is highly visible, and its relative inertness the cause of increasing concern.

Plastic debris is widespread throughout the world. It has, for example, become a common feature of beaches and coastal waters close to densely populated regions of the world, such as Europe and North America and it is widespread along major rivers in industrialised nations (for example, see Figure 11.1). More worrying still, it is an increasing problem in the open sea far away from major navigation routes. In the latter case, plastic pollution tends to be in the form of pellets or beads, as prepared by the polymer manufacturer. These pellets enter the environment in a number of ways, including spillage during cargo loading and losses during transport. Plastic bags are now found throughout the world's oceans, from Spitsbergen 78° North [latitude] to Falklands 51° South [latitude], and are predicted by the British Antarctic Survey to be washing up in Antarctica within a few years.

The effect on marine life of such widespread pollution by polymers is significant. Sea turtles, for example, may mistake plastic pellets for

Figure 11.1   Plastic waste washed up on a beach.
(Photograph: © iStock).

food. Having consumed these pellets, the turtles become too buoyant to dive properly for food and consequently die. Sea birds, too, mistake plastic pellets for food, also with fatal consequences, as the indigestible polymer clogs their intestines. The reduction in the quality of coastal amenities is also serious. People's enjoyment is seriously impaired when beaches are covered with plastic debris, such as discarded poly(ethylene) bags and plastic drinks bottles. Because of the visibility of such debris, there is no doubt that the general public is becoming increasingly reluctant to endure pollution by plastics as part of the inevitable price of progress.

A major difficulty is that so many applications employ polymers as though they were readily disposable, despite the fact that they are, in reality, among the more inert of chemical pollutants. In particular, plastic packaging tends to make a single use of a fabricated polymer which is then thrown away. Even when disposed of responsibly, packaging materials are the source of a serious pollution problem. Municipal authorities dispose of such materials by one of two routes, landfilling and incineration, of which the latter cannot be readily used for chlorinated polymers, such as PVC, because of the additional problems caused by organochlorine species in the atmosphere.

Landfills have been widely used as a means of disposal of waste by municipal authorities. Essentially, the rubbish is simply buried in the ground, though there may be some sort of reclamation process, leaving only a residue for burial in the landfill. However, available land is in short supply, and changing regulations make it more and more difficult to dispose of plastics in this way, partly because of concerns with leaching of additives, and partly because of the general commitment on the part of national governments, including in the UK, to reduce the amount of solid waste going to landfill. Consequently, it is becoming increasingly urgent to deal with the problem at source, rather than once plastics have entered the environment.

As mentioned, packaging materials are among the most widespread of waste plastics. Though they are valuable, for example in the preservation of food as described in Chapter 9, packaging is typically a one-off application, and this has brought these materials under increasing scrutiny. Even when these materials are disposed of responsibly, there are problems with their use. So often, however, they are not disposed of responsibly, and then there are significantly greater problems. Large amounts of used packaging are simply dumped in inappropriate places with little or no regard for the damage they cause to the environment. The volume of such discarded plastic beggars the imagination. For example, studies published in 2014 suggested that there are some 5.25×10^{12} pieces of plastic waste floating in the world's oceans, weighing an estimated 268 940 tons. In 2015, the Marine Conservation Society found an average of 3298 pieces of litter per kilometre of beach surveyed in the UK, of which about 70% were plastics of various kinds. Similar surveys have been carried out for several years, and each year both the number of items and the proportion of plastics increase.

11.3   The Nature of the Problem

The problem of pollution by polymers is not simply technical, but has important social and political aspects. When rightly used, polymers are desirable materials. Their properties make them ideal for a large number of applications. Their persistence in the environment and resistance to degradation are not necessarily the cause of the problem. After all, the great civilisations of the past left behind artifacts that were equally persistent, though these tended to be made of materials such as pottery or glass. The problem with polymers is rather that they are so often used inappropriately. Poly(ethylene) bags and poly(styrene) beakers are designed to be so flimsy that they can be used only once. The materials are simply not appropriate for the fabrication of such 'disposable' items, yet this is what they are used for.

The consequence of this inappropriate materials selection is the squandering of natural resources that characterises the industrialised nations of the world, coupled with a growing problem of waste disposal. Control is certainly necessary to prevent the uncontrolled dumping of these non-degradable polymers in the environment. Arguably control is also needed to prevent these kinds of material being fabricated into barely serviceable items in the first place.

The non-technical nature of the problem becomes apparent when we consider a specific example. For instance, plastic bottles, which are lighter and cheaper than those made from glass, have superseded the traditional material in all sectors of the modern drinks industry. In Britain five billion plastic bottles are used a year, which leads to serious environmental problems. They are difficult to recycle or reuse and expensive to dispose of. They cannot be reused because of the need for sterility. Sterilising is done using high temperatures, which would cause softening or even melting if applied to plastics.

This change to plastic bottles is relatively recent. Previously, many beverages were offered in glass bottles, and some, such as lemonade, were sold in returnable bottles. With the growing role of supermarkets as the main outlet for bottled beverages, the use of returnable bottles has declined. As with so much of this topic, the problem of returnable bottles is clearly not just a technical one. In Germany, France, and Spain supermarkets have encouraged the use of returnable bottles by providing collection points, and because of this customers have happily continued to use and return their empty bottles. As a result, on the continent, glass continues as the material of choice for these bottles.

Plastic bags and their disposal also constitute a significant environmental problems. They are highly likely to end up as litter, not only along beaches, but throughout the countryside, causing loss of amenity and damage to animals and plants. Several countries, including Denmark, Sweden, Germany, Italy and Ireland have implemented policies aimed at reducing this problem. Strategies including charging customers for plastic carrier bags, and taxing their use. This latter approach has been very successful in Ireland, where there has been a significant growth in public awareness of their impact on the environment, and a fall in their use by 90%.

It is instructive to consider the sources of plastics waste. Table 11.1 shows the types of plastics and their use in packaging, whilst Table 11.2 shows the proportion of waste classified by end use. This shows how significant packaging is in generating waste, and causing environmental problems. Applications such as furniture and electrical appliances show more acceptable use of polymers,

Table 11.1 Plastics used in packaging.

Plastic	Use
Poly(ethylene terephthalate)	Drinks bottles, oven-ready tray meals
High density poly(ethylene)	Bottles for milk and washing-up liquid
PVC	Food trays, cling film, squash bottles, Shampoo bottles
Low density poly(ethylene)	Carrier bags, bin liners
Poly(propylene)	Margarine tubs, microwaveable food trays
Poly(styrene)	Yoghurt pots, plastic cutlery, foamed meat/fish trays, hamburger boxes

Table 11.2 Plastic waste generation by end use.

End use	Approximate life/years	% Discards
Packaging	1	100
Transport	5	20
Furniture & housewares	10	10
Electrical appliances	10	10
Building and construction	50	2

because discard rates are low and lifetimes of the products are generally long.

11.4 Polymers and Energy

Since plastics are generally made from hydrocarbon feedstocks they should be recycled to conserve energy. The most effectives energy conservation is to refabricate plastic items, though this is not always technically feasible. Under circumstances where recycling is not a feasible option the use of plastics in waste-derived fuels may be an acceptable conservation measure.

Burning of plastics is not a favourable option, given the widespread concern over the global increase in carbon dioxide. Nonetheless these materials are potentially useful as fuels, and burying them simply wastes their potential in this respect. On the other hand, burning plastics may release not only CO_2 but also trace amounts of other pollutants, all of which are undesirable.

11.5 Recycling of Polymers

Recovery and reuse of synthetic polymers is by far the most acceptable way of dealing with the problem of waste. Items can rarely be reused as such but instead the polymers from which they are fabricated can be recycled. In principle, polymers can be recycled without any significant loss of their properties. For example, polycarbonate bottles can be recycled into automobile bumpers and then into articles for high performance housing.

The current problem with recycling is economics. It is not generally cost effective to recycle the common high-tonnage polymers such as poly(ethylene), poly(styrene), and other commodity polymers because the recovered material has a low intrinsic value. This stems in part from the fact that the original polymers depend for their popularity on their ease of manufacture and consequent low cost. The challenge, then, is to learn how to organise the collection of waste material, followed by separation and reprocessing, in a way that is profitable for the organisations involved.

In America there are promising signs for certain polymers. For example, poly(ethylene terephthalate) drinks bottles can be cleaned and recycled to give an acceptable grade of PET resin in a process that is economically viable. The recycled polymer is used as carpet fibre, furniture stuffing, or insulation. Waste nylon can also be recycled profitably.

The quantity of these materials is relatively small compared with the amount of waste high-density poly(ethylene) produced each year. Containers made from HDPE are widely used for detergents, oil, and antifreeze, and enormous amounts of material are used in disposable applications annually. In principle recycled poly(ethylene) could be used for drain pipes, flower pots, dustbins, and plastic crates. The problem remains, however, that economics do not favour recycling of these polymers and in the absence of Government intervention little or nothing can be done to alter commercial attitudes towards recycling.

One technical difficulty that does beset recycling is that in many applications a variety of polymers are employed together in a complex way. It therefore becomes essential to distinguish between the various types of polymer in order to separate them. One system proposed (but not yet introduced anywhere in the world) is for the individual polymer components of complex articles such as automobiles to be identified using computer-scannable bar codes on each individual polymer component.

Recycling can include chemical recycling. This involves plastics waste being processed chemically, either by "cracking" or depolymerisation, to recover material and process it into a form that can be used industrially. Plastics waste is usually mixed and there are particular issues to be considered if the mixture contains substantial amounts of PVC. This is because of the importance of recycling the chlorine without releasing it to the atmosphere, and because PVC typically contains metal-based stabilisers, and these generate solid waste that needs to disposed of with care.

A number of cracking/depolymerisation processes are currently operating commercially. One of these is the Texaco gasification process, which has been operating since the mid 1990s. In this process, there are two steps, an initial liquefaction step followed by treatment in an entrained bed gasifier. The liquefaction step involves heating the plastic scrap in such a way that partial depolymerisation occurs, generating a heavy oil and some gas fractions. Some of the gas is recycled as fuel for the process.

The heavy oil is filtered to remove inorganic particles, then injected into the entrained gasifier. Gasification is carried out in the presence of oxygen and steam at 1200–1500 °C, and after cleaning of HCl (from PVC) and/or HF (from PTFE) a dry synthesis gas is produced. This consists mainly of CO and H_2 with smaller amounts of CH_4, CO_2, H_2O and some inert gases. The removal of chlorine in this process is achieved by washing the raw gas in solutions of ammonia which causes the chlorine to be converted to NH_4Cl, which can be sold commercially.

11.6   Degradable Polymers

An alternative tactic to deal with the problem of polymer wastes is to make polymers degradable. The difficulty with this approach is that in making synthetic polymers degradable one of the greatest assets of these materials, namely their durability, may be eliminated. There is also the possibility that degradation may give rise to reaction products that may themselves cause environmental problems. For example toxic wastes from degrading polymers have caused problems of water pollution in the vicinity of landfills.

Not all synthetic materials are worth recovering. For example plastic grocery bags, six-pack drinks can rings, fishing nets, and marine ropes are generally difficult to recover. If the degradation products are environmentally benign, then making these items of degradable polymers could solve the problem of waste.

Several approaches exist to designing degradable polymers but all act slowly. Degradable polymers must have a reasonable lifetime in their original service; plastic grocery bags cannot begin to degrade to any extent from the moment they are produced or else they would be unacceptable for the purpose for which they are designed. Hence, degradable polymers tend to be designed to have the degradation triggered, for example by exposure to ultraviolet light.

One approach under active consideration as a means of reducing the problem of plastics waste is to develop polymers that are biodegradable. The degradation processes are carried out by micro-organisms, either bacteria or fungi, and are generally complex.

Natural macromolecules such as cellulose or proteins are generally degraded in biological systems by hydrolysis followed by oxidation.

In designing synthetic polymers that are capable of being degraded biologically use is made of functional groups that are themselves susceptible to enzymic hydrolysis and oxidation. Polymers which fulfil this requirement include aliphatic polyesters, polyurethanes, poly(vinyl alcohol), and poly(vinyl ethanoate). These often find application in medical areas, such as absorbable implants and sutures or as controlled-release matrices for drug delivery. Similar materials are now being developed for use in packaging applications which are conventionally the province of non-degradable polymers.

Soil burial is widely used as the method of testing susceptibility to degradation. It closely mimics the conditions of waste disposal used for plastics but it is often difficult to reproduce results obtained because of absence of control over either the climate at the test site or the variety of micro-organisms involved in the degradation. Soil burial is thus used to provide qualitative indications of biodegradability, with more controlled laboratory work with cultured micro-organisms being used to obtain more quantitative detail.

Such studies have shown that it is the chemical structure and composition that determine whether or not synthetic polymers are biodegradable. On the other hand, the precise rate at which a synthetic polymer will degrade is determined by the specific morphology of the article into which the polymer has been fabricated.

11.7   Green Polymerisation

This is a generic term for an area that encompasses activities ranging from polymerisation of monomers from renewable resources to polymerisation reactions carried out under more environmentally friendly conditions than conventional methods.

In principle, there are several possible types of monomer available from renewable resources. These include plant oils, sugars and polysaccherides, glycerol, lignin and its derivatives, and mono-terpenes. The latter can also be used as solvents for polymerisation.

To date, however, work on these polymers is in its infancy, and none of the products has achieved any widespread use. Indeed, for a variety of reasons, it is difficult to see such polymers ever rivalling the major tonnage polymers such as polyethylene or polypropylene. On the other hand, there is every prospect of them succeeding and becoming established as materials of choice for niche applications.

One interesting recent development is the use of ionic liquids as polymerisation solvents. Ionic liquids are salts of large organic ions, typically of nitrogen, associated with large counterions such as tetrafluoroborate, BF_4^-. These ions are bulky and have diffuse charges, so that the inter-ionic forces between them are relatively low, and the substance is a liquid, albeit a viscous one, at room temperature. The ionic nature of these substances means that they are almost completely involatile, with no vapour pressure at all. This gives them considerable potential as "green" solvents as they do not pollute the environment and they are typically fairly easy to recycle.

An example of the use ionic liquids in polymerisation is the deployment of imidazolium salts of tetrafluoroborate to polymerise methyl methacrylate. An example of these salts is 1,3 dibutyl imidazolium teatrafluoroborate, which gives poly(methyl methacrylate) with high molar mass. The rate of polymerisation is much higher than with conventional solvents, or even in bulk, an effect that arises from the fact that ionic liquids reduce the activation energy of both the propagation and termination reactions.

Although it is early days for this particular use of ionic liquids, the results obtained so far suggest that this approach to polymerisation will grow to become significantly in the future.

11.8   The Future

The problems posed by polymers in the environment are not solely, or even mainly, technical. Instead, they concern both politics and the behaviour of individuals within society. Litter, for example, is a social problem; if specific individuals did not choose selfishly to discard their rubbish where they stood in the street, at the picnic site, or on the beach, there would be no litter. However, people do not behave for the common good but, sadly, generally favour the selfish option and such antisocial behaviour must be taken account of in any scheme for reducing the impact of polymers on the environment.

In fact legislation is the route that is being taken to address many aspects of these problems. The late 1980s saw many steps taken concerning plastics and their entry into the environment. For example, in America, there was the US Plastic Pollution Research and Control Act, which prohibits the dumping of any plastic at sea by any US ship throughout the world, and by any ship, regardless of nationality, with 200 miles of the US coast. There was a similar measure adopted internationally in an agreement called MARPOL, Annex V of which outlines steps to prevent the dumping of polymers into the marine environment. This latter prohibition extends to synthetic ropes and fishing nets, as well as to the array of plastic litter discussed previously.

On land, too, there are changes in the way plastics are disposed of. In the UK some local authorities are actively promoting recycling of plastic bottles and poly(styrene) containers, for example by schemes where local residents place recyclable materials in designated boxes for refuse collectors. These boxes are taken to a Materials Reclamation Facility for final sorting of the plastics waste.

Within Europe there is legislation aimed at dealing with problems arising from the use of plastic packaging. This follows from the 1994 EU Directive on Plastics and Packaging Waste 94/62/EC, the so-called Packaging Directive, which requires the setting and monitoring of targets for recycling by national governments. Within Europe, it has led to some 34.7% of plastics being recycled in 2012, with the aim of increasing this proportion in future years.

Certainly polymers are the building blocks for materials with a desirable range of properties that enhance the quality of modern life. They cannot be uninvented, nor is it necessary that they should be. However, the exploitation of polymers will have to be carried out more thoughtfully in the future than it has been in the past. Moreover, the social and political questions concerning their use and disposal will remain pressing well into the future.

Further Reading

J. P. Greene, *Sustainable Plastics*, John Wiley and Sons, Chichester, 2014.

Green Polymerization Methods: Renewable Starting Materials, Catalysis and Waste Reduction, ed. R. T. Mathers and M. A. R. Meier, Wiley, Chichester, 2011.

Bibliography

Comprehensive Polymer Science, ed. G. Allen and J. C. Bevington, Pergamon Press, Oxford, 1989, vol. 1–7.

J. A. Brydon, *Plastics Materials*, Butterworth-Heinemann, London, 7th edn, 1999.

C. E. Carraher, *Introduction to Polymer Chemistry*, CRC Press, Boca Raton, Florida, USA, 3rd edn, 2013.

J. M. G. Cowie and V. Arrighi, *Polymers – Chemistry and Physics of Modern Materials*, CRC Press, Boca Raton, Florida, USA, 3rd edn, 2007.

P. C. Hiemenz and T. P. Lodge, *Polymer Chemistry*, CRC Press, Boca Raton, Florida, USA, 2nd edn, 2007.

S. Koltzenburg, M. Maskos, O. Nuyken and R. Mulhaupt, *Polymer Chemistry*, Springer, 2016.

G. Odian, *Principles of Polymerization*, John Wiley and Sons, New York, 4th edn, 2004.

P. C. Painter and M. M. Coleman, *Essential Polymer Science and Engineering*, DESTech Publications, Lancaster, Pennsylvania, USA, 2009.

R. J. Young and P. A. Lovell, *Introduction to Polymers*, Chapman and Hall, London, 3rd edn, 2011.

The Chemistry of Polymers, 5th Edition
By John W. Nicholson
© John W. Nicholson 2017
Published by the Royal Society of Chemistry, www.rsc.org

Subject Index

acetone, 14
acrylic polymers, 9
acrylonitrile, 10
acrylonitrile-butadiene-styrene
  (ABS), 145
addition polymers, 5
adhesives, 15
adipic acid, 13
alkyd resins, 66
allophanates, 74
alternating copolymerisation, 46
amino acids, 23
ω-aminodecanoic acid, 13
amino resins, 16–17
ammoniacal cupric hydroxide, 20
anionic mechanisms, 37
antimony–halogen synergism, 140
antioxidants, 143, 144
arborols, 152
artificial weathering, 142
asymmetric carbon atoms, 49
atmospheric oxygen, 66
atomic force microscopy (AFM),
  165
atom transfer radical
  polymerisation (ATRP), 41–42
AT&T Bell, 177
autoacceleration, 32–33
Avogadro's number, 107
azobis-isobutyronitrile (AIBN), 27

Baekeland, L., 15
bakelite, 15

barium laurate (PVC stabilizer), 12
Bayer, O., 72
Benoit, H., 107
benzene, 2
benzoyl peroxide, 27
bifunctional acid, 65
Billmeyer, F.W., 48, 95
biological degradation, 145–146
p,p'-biphenol, 183
bisphenol A, 14
Boltzmann equation, 81
brittle fracture, 112–115
bulk polymerisation, 33–34
butadiene, 10, 11, 39
butadiene isoprene, 49
but-1-ene, 39
butyl rubber, 51

cadmium laurate (PVC stabilizer),
  12
calcium silicate, 182
caoutchouc, 21
caprolactam, 13
carbenium ion, 37
carbon, 2
carbon atoms, 177
carbon–carbon bonds, 177
carbon disulfide, 20
Carothers, W., 4, 5, 12, 25
cascade molecules, 152
Castan, P., 75
cationic mechanisms, 37
cellophane, 22

cellulose, 20–21
chain polymerisation, 25, 26
chain polymerisation, kinetics of, 30–32
char, 137
chlor-alkali, 174, 175
chlorine, 2
cold drawing, 123–124
colligative properties, 95–97
combustion behaviour, 138–139
condensation polymerisation, 5, 12
configuration, 48
conformation, 48
contour length, 85
co-ordination chemistry, 158
co-ordination polymer, 191–192
copolymerisation, 45–47
cosmetics, 21
creep relaxation, 121–122
cresol, 15, 16
crosslinks, 64
crystalline melting point, 52–53, 62
crystallisation, 52

de Gennes, P., 86
degradable polymers, 200–201
degree of polymerisation, 1
dehydrochlorination (of PVC), 12
dendrimer, 152
dendrimer applications, 166–168
dendrimer characterization, 161–166
dendrimer conformations, 159–160
dendrimer micelles, 160–161
dendrimer shapes, 159–160
dendrons, 152
diaminodiphenylmethane (DDM), 76
dielectric thermal analysis, DETA, 59

differential scanning calorimetry, DSC, 58
4,4-difluorobenzophenone, 19
dilute solution, 83–84
diphenyldimethoxysilane, 39
diphenylmethane, 4,4′-di-isocyanate, 18
disproportionation, 28
dissolution, 77–78
dodecyl lactam, 13
Dow Chemical Company, 156, 158
drug delivery, 166–167, 187–190
Du Pont, 4, 12
dynamic mechanical thermal analysis, DMTA, 59
dynamic tests, 124–126

ebonite, 22
electromagnetic radiation, 26
electronically conducting polymers, 176–178
emulsion polymerisation, 35–36
end group analysis, 104–105
energy, 198
entanglements, 86
entropy of mixing, 81–83
environmental stress cracking [poly(propylene)], 9
epoxy resins, 14, 15, 74–76, 179
ethanal, 1
ethene/ethylene, 6
ethylene, 39
excess chemical potential, 83

Farbenindust, I.G., 72
Fenimore, C.P., 138
fires, 136–137, 139–140
flame retardants, 139, 140
Flory equation, 100
Flory–Huggins theory, 82
Flory, P.J., 5, 77, 82
fluorine, 2, 17

food-contact polymers, 186
food packaging, 185–187
formaldehyde, 15, 16, 17
formic acid, 16
fracture toughness, 119
free radicals, 27, 29

Gaussian distribution, 83
gel permeation chromatography,
   GPC, 77, 94, 105–108, 162
General Bakelite Company, 15
gene therapy, 167
Gibbs equation, 128
glass transition temperature,
   54–57, 58–61, 62
glucose, 20, 21, 24
glycol, 70
Greenlee, S.O., 75
green polymerisation, 201–202
Griffith, A.A., 117
Griffith crack equation, 118

hafnium, 39
Harries, C.D., 3
head-to-tail configuration, 30
Hevea brasiliensis, 21
hexamethylene di-isocyanate, 18
hexamethylenetetramine
   (HMT), 67
high-performance liquid
   chromatography, HPLC, 162
high pressure process, 7
high-viscosity polymer solutions,
   89–90
homo-disperse, 93
Hooke's law, 120
Huggins, M.L., 82
hydrogen, 2
hydrogen abstraction, 28
hydrogen-bonding, 78, 80
hydrogen bonds, 2
hydrogen chloride, 12

hydroquinone, 19
p-hydroxybenzoic acid, 183

industrial catalysts, 167–168
influence of surfaces, 116–119
initiation, 26–27
inorganic polymers, 180–183
interpenetrating polymer
   network (IPN), 178–180
iodobenzene, 168
ionic interactions, 158
ionomers, 173–175
isocyanate group, 18
iso-quinoline-bromine complex,
   32
isotactic configurations, 49

Kelvin model, 120
kinetics of chain polymerisation,
   30–32
Krigbaum, W.R., 83, 86

landfills, 195
lead soaps (PVC stabilisers), 12
Lewis acids, 14
Lewis bases, 14
light emitting diode (LED), 150
light scattering, 97–101
limiting oxygen index (LOI),
   138, 139
liquid ammonia, 38
lithium–polymer batteries,
   190–191
living polymers, 38, 39–41
L-lactic acid, 189
loop process, 36–37

macromolecule, 2–4
magnetic resonance imaging,
   MRI, 166
maleic acid, 70
Mark–Houwink equation, 103, 104

mass spectroscopy, 161
matrix-assisted laser desorption/
    ionization (MALDI) mass
    spectrometry, 108–110, 161
Maxwell model, 120
mechanical properties, 54, 133–135
medicine, 170–173
megamers, 168–169
melamine, 16, 17
metastyrene, 3, 10
methanol, 9
methyl group, 24
methylol units, 16
molar masses, 95–97
molecular motion, reptation
    model of, 86–87
mono-disperse, 93
monomer reactivity ratios, 47
monomers, 26
monomer units arrangement,
    29–30
Morgan, Sir Gilbert T., 15
Mulder, G.J., 22

nafions, 174
naphthalene diazoquinone
    sulfonate (NDS), 149
Natta, G., 7, 8
natural rubber, 21–22
nature of problem, 196–198
N-cyclohexyl-2-pyrrolidone, 19
Newtonian liquid, 124
nitric acid, 16
NMR spectroscopy, 163
non-crystalline polymers, 51
non-rubbery polymers, 52
number average molar mass, 93
nylon, 12–14
nylon 6,6, 13, 43, 44
nylon 6,10, 13
nylon 12, 13
nylons, 171

organic dendrimers, 155–158
orientation, 52
osmotic pressure, 95
Ovchinnikov, A.A., 177
oxirane ring, 14
oxygen, 2, 7, 10, 19
oxygen–nitrogen mixture, 138

phenol, 15–16
phenol-formaldehyde polymers,
    15–16
phenol–formaldehyde resins,
    67–69
Phillips process, 7
phosphorus, 2
photo-oxidation, 143–145
photo-oxidative degradation, 142
phthalic acid, 70
plasticisers, 57–58
pollution, 193–196
poly(acetylene), 176, 177
poly(acrylamide), 88
poly(acrylic acid), 9, 88, 89, 188
poly(alkenes), 145
poly(butadiene), 92
poly(*cis*-1,4-isoprene), 49
poly(dimethoxysilane), 92
poly(ethylene), 34, 51, 92, 173,
    176, 188
poly(ethylene glycol), 92, 188
poly(ethylene terephthalate), 52
poly(glycolic acid), 171, 188
poly(hydroxyethyl methacrylate),
    171
poly(isobutylene), 51
poly(isoprene), 22
poly(lactic acid), 188
poly(methacrylic acid), 89
poly(methyl methacrylate), 28, 34,
    49, 54, 55, 113, 117, 171, 188
poly(methylmethacrylate), 138
poly(*N*-isopropyl acrylamide), 188

poly(oxymethylene), 138
poly(*p*-phenylene), 176, 177
poly(*p*-phenylene sulfide), 177
poly(*p*-phenylene vinylene), 151
poly(phenylenetetraphthalamide),
    183
poly(phosphazenes), 180
poly(propylene), 52, 188
poly(pyrrole), 176, 177
poly(stryrene), 145
poly(styrene), 10–11, 34, 53, 54,
    92, 138
poly(vinyl acetate), 188
poly(vinyl alcohol), 1, 88, 188
poly(vinyl ethanoate), 1, 88
poly(vinylidene chloride), 52
polyamidoamines (PAMAMs), 156
polyblends, 53–54
polydispersity, 93
polymer crystallinity, 50–52
polymer degradation, 147–150
polymer light emitting diodes,
    150–151
polymer liquid crystals,
    183–185
polymer melts, 90–92
polymer–solvent interaction, 96
polymer stereochemistry,
    48–50, 61
poly(ether ether ketone), PEEK,
    18–19
poly(ethylene glycol), PEG, 189
polypeptide, 23
poly(L-lactic acid), PLLA, 189
poly(methyl methacrylate),
    PMMA, 9–10
poly(tetrafluoroethylene), PTFE, 17
poly(vinyl chloride), PVC, 1, 11–12,
    35, 138, 171, 188
polystyrene, 28
polyurethanes, 18, 72–74, 171, 188
porphyrin, 168

potassium persulfate, 35
propagation, 27–28
1,2-propanediol, 70
propionic acid, 24
propylene, 39
1,2-propylene glycol
    (1,2-propanediol), 70
proteins, 22–23

radius of gyration, 85–86
random coil conformations, 84
Raoult, F.M., 80
Raoult's law, 80–81
rate constant, 31
γ-ray, 27
Rayleigh ratio, 98
recycling of polymers, 198–200
reinforced polymers, 130–133
relative molar mass, 102–104
reversible addition
    fragmentation (chain) transfer,
    or RAFT, 39–41
ring opening methathesis
    polymerisation, ROMP, 42–43
root-mean-square end-to-end
    distance, 85
rubber, 21–22
rubberlike elasticity, 127–130

self-diffusion coefficient, 87
shapes of polymer molecules,
    84–86
silicon, 19–20
silicones, 20–21
Simon, E., 10
simple liquid mixtures, 80–81
small angle neutron scattering,
    SANS, 164–165
sodium hydroxide, 16
solubility parameter, 79
solubility parameters, 78–80
solution polymerisation, 34–35

Standard Oil (Indiana) process, 7
Staudinger, H., 3, 4, 21
step polymerisation, 25, 43–44
step polymerisation with polyfunctional monomers, 44–45
step reactions, 5, 6
strain, 112
strength, 115–116
stress, 112
stress relaxation, 121–122
stress–strain curves, 113–114
Strutt, J. (Lord Rayleigh), 97
styrene, 3, 39
styrene-butadiene rubber, SBR, 10
sulfur, 2, 12, 22
supramolecular dendrimers, 158
suspension polymerisation, 35
syndiotactic configurations, 49
synthetic approach, 153–154
synthetic metal, 176

tail-to-tail configurations, 30
tandard linear solid, 120
terephthalic acid, 183
termination, 28
tetracar-boxylic acid moieties, 158
Thakur, M.K., 177
thermal properties, 54
thermal transitions, 62–63
thermoplastics, 4, 5, 6, 137
thermoset, 4, 5, 6, 15
time/temperature relationship, 126–127
tin di-iso-octylthioglycollate (PVC stabiliser), 12
titanium tetrachloride, 7
titration, 164

tolylene di-isocyanate, 18
tough behaviour, 65
tough fracture, 112–115
trans vs. gauche arrangements, 48
triethylaluminium, 39

unimolecular micelles, 160
unsaturated polyester resins, 69–72
urea, 16
urethane, 18

vapour phase osmometry, 96–97
vinyl alcohol, 1
vinyl benzene, 10
vinyl monomers, 29
viscoelasticity, 119–121
Voigt model, 120
vulcanisation, 22

water, 5, 9, 16
water-soluble polymers, 87–89
weathering of polymers, 141–143
weight-average molar mass, 93

X-ray diffraction, 163–164
xylenols, 15

yield strength, 114
Young's modulus, 59, 112

Zacheriasen, W.H., 180–181
z-average molar mass, 94
Ziegler, K., 7
Ziegler–Natta catalysts, 38
Ziegler–Natta polymerisation, 177
Ziegler process, 7
zinc dust, 11
zinc oxide, 22
zirconium, 39